Handheld Computers in Schools and Media Centers

Ann Bell

Professional Development Resources for K-12
Library Media and Technology Specialists

Library of Congress Cataloging-in-Publication Data

Bell, Ann, 1945-
　Handheld computers in schools and media centers / Ann Bell.
　　p. cm.
　　Includes bibliographical references and index.
　　ISBN 1-58683-212-3 (pbk.)
 1. Educational technology. 2. Wireless communication systems. 3. Mobile communication systems. 4. School libraries--Automation. I. Title.
　LB1028.3.B45 2007
　371.33'4--dc22
　　　　　　　　　　　　　　　2006025691

Cynthia Anderson: Acquisitions Editor
Carol Simpson: Editorial Director
Judi Repman: Consulting Editor

Published by Linworth Publishing, Inc.
480 East Wilson Bridge Road, Suite L
Worthington, Ohio 43085

The term iPod™ is a trademark of Apple Computer, Inc.
Apple Computer, Inc., does not endorse the opinions
within this book, and there is no relationship between
Linworth Publishing, Inc. and Apple Computer, Inc.

Copyright © 2007 by Linworth Publishing, Inc.

All rights reserved. No part of this book may be
electronically reproduced, transmitted, or recorded
without written permission from the publisher.

ISBN: 1-58683-212-3

5 4 3 2 1

table of contents

Table of Figures	vii
About the Author	ix
Acknowledgments	ix
Introduction	xi
Chapter 1: Using Handheld Devices to Meet National and State Standards	1
Standards for the English Language Arts	2
The National Education Technology Plan	3
The National Educational Technology Standards for Students (NETS-S)	4
The National Educational Technology Standards for Teachers (NETS-T)	6
The Educational Technology Standards for Administrators	7
Information Literacy Standards	9
Summary and Challenge	11
Chapter 2: Selecting Hardware for eBooks, eAudio, eVideo, and Podcasting	13
Types of Handheld Devices	14
Selection Consideration Factors	15
Peripherals	19
Summary and Challenge	20
Chapter 3: Selecting Software for the Handheld Device	21
Synchronizing Handheld Computers to Desktop or Laptop Computers	22
Document Preparation Software	22
Printing from Handheld Devices	22
Standards for Digital Media	23
Audio Standards for Digital Talking Books	23
eBook Readers for Handheld Devices	24
Media Players for Handheld Devices	26
Subject Specific Software	30
Summary and Challenge	30
Chapter 4: Locating and Downloading Online eBooks	31
History of eBooks	32
Hazards of eBooks	32

Educational Advantages for eBooks	33
Transferring eBooks from Desktop to Handheld Devices	33
eBook Formats	34
Sources for eBooks	36
College and University Electronic Book Collections	39
eBooks for People with Disabilities	39
Specialized eBook Search Engines	40
Summary and Challenge	40

Chapter 5: Accessing Web Sites on Handheld Computers — 41

Mobile Browsing	42
Using Mobile Favorites	43
Web Clipping Software	45
AvantGo	46
Streaming Digital Media via the Web from a Desktop to a Handheld Device	46
Summary and Challenge	47

Chapter 6: Writing eBooks and Notes — 49

Curricular Uses of Student Produced eBooks	50
eBook Writers	50
Creating eBooks for the iPod	54
Summary and Challenge	54

Chapter 7: Circulating eBooks and eAudiobooks — 55

Distributing Applications and Documents	56
Advantages of Circulating Digital Media in School Libraries	56
Locating eBook Managing Sources	57
Creating and Managing an eAudio Library	59
Creating a School eLibrary	60
Summary and Challenge	60

Chapter 8: Utilizing and Preparing eAudio — 61

Educational Value of Audiobooks	62
Selecting Audiobooks	63
Audiobook Industry	63
Sources of Audiobooks	64
Ripping Audiobooks	65
Direct Audio on a Handheld Device	65
Converting Text to Audiobooks	66
Learning Foreign Languages	67
For the Disabled	67
Online Radio	67
Summary and Challenge	68

Chapter 9: Utilizing and Preparing Podcasts   69
    What is Podcasting?   69
    History of Podcasting   70
    Educational Uses of Podcasts   71
    Legal and Copyright Issues in Podcasting   73
    Hardware Needed to Listen and Produce a Podcast   74
    Software Needed to Listen and Produce a Podcast   76
    Locating and Listening to Podcasts   80
    Recording a Podcast   82
    Hosting and Publishing a Podcast   85
    Summary and Challenge   86

Chapter 10: Locating and Downloading eVideos and Vodcasts   87
    History of Vodcasts   88
    Sources for Educational eVideos and Vodcasts   88
    Selecting Software and Media Players for Video Playback   90
    Converting Video Formats to Play on a Handheld Device   92
    Shooting Video for the Micro Screen   94
    Preparing a Vodcast   94
    Summary and Challenge   95

Chapter 11: Digital Media Copyright Issues   97
    Fair Use Guidelines   98
    Berne Convention   98
    Visual Artists Rights Act   98
    Fair Use Guidelines for Educational Multimedia   99
    The Digital Millennium Copyright Act   99
    Digital Performance Rights in Sound Recordings Act   100
    TEACH Act   100
    Software Downloading Regulations   100
    Summary and Challenge   101

Chapter 12: Incorporating eBooks, eAudio, eVideo, and Podcasts into the Curriculum   103
    Collaboration   104
    Teacher-Produced Assignments   104
    Reading Strategies with eBooks   104
    Cross-Curricular Uses of eBooks   106
    eBooks in Language Arts   106
    eBooks in Foreign Languages   107
    eBooks in Science Curriculum   107
    eBooks in Social Studies Curriculum   107
    Using eAudio within the Curriculum   107
    Mobile Devices for Students with Special Needs   108

Using eVideo within the Curriculum	108
Using Podcasting within the Curriculum	109
Summary and Challenge	110

Chapter 13: Record Keeping on Handheld Devices — 111

Inventory Control	112
Schedule and Appointment Reminder	113
Track Student Progress	113
Recording and Tabulating Grades	114
Attendance Tracker	114
Lesson Planner	114
Student Information Systems (SIS)	115
Student Organizational Software	115
Managing Classroom Sets of Handheld Computers	115
Managing Computer Networks with Handheld Devices	116
Summary and Challenge	117

Appendix A: eBook, eAudio, and eVideo Formats — 119

Sources Consulted — 121

Glossary — 125

Index — 129

table of figures

Chapter 1
Figure 1:1 – Vision Guide ... 2
Figure 1:2 – Innovative Budgeting ... 3
Figure 1:3 – Digital Content ... 4
Figure 1:4 – Indicator 4 – Sources ... 10
Figure 1:5 – Indicator 4 – Information ... 10
Figure 1:6 – Indicators 1 and 2 ... 10
Figure 1:7 – Indicators 2 and 3 ... 11

Chapter 2
Figure 2:1 – CompactFlash Card ... 18
Figure 2:2 – MultiMediaCard ... 18
Figure 2:3 – SecureDigital Card ... 18
Figure 2:4 – Sony Memory Stick ... 18
Figure 2:5 – External Keyboard ... 19
Figure 2:6 – Virtual Keyboard ... 19

Chapter 3
Table 3:1 – eBook Reader Comparison Chart ... 25
Figure 3:1 – MPEG ... 26

Chapter 4
Figure 4:1 – Adobe Reader on a Handheld Computer ... 34
Figure 4:2 – MobiPocket Reader on a Handheld Computer ... 35
Figure 4:3 – Microsoft Reader on a Handheld Computer ... 35
Figure 4:4 – eReader Pro on a Handheld Computer ... 36

Chapter 5
Figure 5:1 – Screenshot of Explorer Toolbar ... 43
Figure 5:2 – Screenshot of Pop-up Screen ... 44
Figure 5:3 – Screenshot of Next Pop-up Screen ... 44
Figure 5:4 – Screenshot of Organize Box ... 45

Chapter 6
Figure 6:1 – Screenshot of Microsoft Reader Plug-in ... 51
Figure 6:2 – Screenshot of ReaderWorks Program ... 52
Figure 6:3 – Screenshot of DropBook Program ... 52
Figure 6:4 – Screenshot of Palm eBook Studio Program ... 52
Figure 6:5 – Screenshot of MobiPocket Companion Program ... 53

Chapter 8
    Figure 8:1 – Griffin iTalk   65
    Figure 8:2 – MicroMemo   66

Chapter 9
    Figure 9:1 – Full-cuff Headset   75
    Figure 9:2 – Sound Mixer   75
    Figure 9:3 – RSS Orange Box   77
    Figure 9:4 – Podcast Signal   78
    Figure 9:5 – Screenshot Windows Media Player Library   81
    Figure 9:6 – Workflow in Recording a Podcast with GarageBand   83

Chapter 10
    Figure 10:1 – PocketDISH   90
    Figure 10:2 – Video iPod   91

Chapter 11
    Table 11:1 – Fair Use Guidelines for Educational Media   99

Chapter 12
    Figure 12:1 – Annotation Page of Microsoft Reader on Handheld Device   106
    Figure 12:2 – Dictionary Page of Microsoft Reader on Handheld Device   106

Chapter 13
    Figure 13:1 – Symbol Handheld Computer   112
    Figure 13:2 – Screenshot of BookBag Software   112
    Figure 13:3 – Screenshot of AudioList Software   112
    Figure 13:4 – Screenshot of VideoList Software   113

acknowledgments

I would like to thank Joan Vandervelde, Online Professional Development Coordinator for the University of Wisconsin-Stout, for her encouragement and support in developing the online class, *Learning Applications for the iPod® and Handheld Computers,* that led to this book. Without her support in providing quality professional development online and her support of educational technology, this book would not have happened.

about the author

Ann Bell is an Adjunct Online Professional Development Instructor for the University of Wisconsin-Stout. She has written and currently instructs the courses *Applying Handheld Computers to the Curriculum* and *Digital Media and Visual Literacy.* Mrs. Bell has been the high school media specialist at Camanche High School, Camanche, Iowa since 1996. Before coming to Camanche, she was a media specialist in Montana, Oregon, and Guam.

Ann received her B.A. degree from the University of Northern Iowa, a library media endorsement from the University of Montana, and a Master of Library and Information Studies from the University of Hawaii. She has completed postgraduate work at Drake University and the University of Northern Iowa.

Ann was the recipient of the 2001 Information Technology Pathfinder Award from the American Association of School Librarians and the Follett Software Company and the 2002 recipient of the Iowa Educational Media Association Lamplighter Award. She was featured in the December 2001 issue of *Teacher Librarian: The Journal for School Library Professionals* and has been published in *Library Media Connection* (Linworth Publishing, Inc.).

Ann has published a series of eight novels for Heartsong Presents Book Club. Barbour Publishing combined the first four books into *Montana,* which was on the Christian Booksellers' Association Bestseller List in 2000. *Montana Skies,* the combined last four books of the series, was published in May 2002. The novel, *Mended Wheels,* was released in July 2002. In 2005, Gale-Thorndike Press released the first four books of the series in a large print edition.

introduction

The integration of a multimedia-educational mobile environment into the classroom increases students' visual and media literacy while it improves standardized test scores. As the use of handheld devices increases within the schools, the ability to collaborate with peers and share that work far beyond the confines of their local community can expand the students' self-confidence and creativity as never before.

State and national standards for student learning emphasize the importance of a student-centered approach to instruction. With each student using an education center stored within his personal handheld devices, the potential of engaging that student in manipulating text and media at any time anywhere can bring amazing results. The desire to use a mobile device will assist students in taking advantage of his multiple intelligences and the flexibility of his time allocated for learning.

In recent months, the power in handheld devices has multiplied and educators are now able to use handheld devices for the additional learning tools of eBooks, eAudio, eVideo, and podcasting. With the advent of the iPod®, student use of handheld devices for entertainment has exploded. With a little direction, these entertainment centers can also become an education center as well.

Library media specialists are leaders in their schools in evaluating hardware and software, locating educationally sound resources, and applying those resources to the curriculum. Librarians are often the first to introduce new resources to the students and to provide professional development for the faculty. The book *Handheld Computers in Schools and Media Centers* was developed to help media specialists, and teachers in general, add another layer of tools to help meet the needs of their students.

After several years of using my handheld computer strictly as a personal information manager so my date book, addresses, tasks, and notes were always available, I realized that I had a lot of power in the palm of my hand that I was not using. I began to question what would happen if our students each had a handheld device to not only help organize their schoolwork, but to use as an educational tool. To help my students take full advantage of the added power and features in handheld devices, I searched the Internet, subscribed to many Really Simple Syndication (RSS) Web feeds, and read numerous books and professional journals to discover ways to increase the use of the handheld computers within the curriculum and to help meet national and state standards. With that information, I developed for the University of Northern Iowa an online professional development course titled *Applying Handheld Computers (PDAs) within the Curriculum*. A year later, the entire professional development program transferred to the University of Wisconsin-Stout. Since that time, the course has evolved into the current course *Learning Applications for the iPod® and Handheld Computers* <www.uwstout.edu/soe/profdev/handhelds/>.

While teaching professional development classes of busy professionals, I found several books that covered classroom activities using handheld computers, but information on handheld computers as a source for eBooks, eAudio, eVideo, and podcasting was extremely limited. To meet that need I compiled my research, spent hours experimenting with handheld devices with various formats and operating systems, and finally this book took form.

Educational technology can add a great deal of excitement and interest into the curriculum, but if that excitement does not transfer into meeting national, state, and local educational standards it will be of little value. Chapter 1, "Using Handheld Devices to Meet National and State Standards," provides detailed analysis of specific standards that could be met by students, teachers, media specialists, and administrators if handheld devices were properly applied to the educational environment.

When considering the possibilities of adapting handheld devices into the curriculum, one of the first decisions to be made is the selection of the hardware. The World Wide Web contains an overwhelming number of possibilities. Chapter 2, "Selecting Hardware for eBooks, eAudio, eVideo, and Podcasting," helps take the mystery and frustration from the selection process.

Chapter 3, "Selecting Software for the Handheld Device," provides a wide overview of the free or inexpensive software for handheld devices that are available for download or purchase. A strong emphasis is on the various types of media players necessary for different audio and video formats.

One of the first uses of handheld devices was the use of text in a portable format. Chapter 4, "Locating and Downloading Online eBooks," covers the history, the hazards, and the educational advantages of eBooks. How to locate free or inexpensive eBooks, the formats available, and how to transfer an eBook to a mobile device is explained in depth.

Chapter 5, "Accessing Web Sites on Handheld Computers," expands the use of digital content even further. Mobile browsing, mobile favorites, Web clipping software, and the selection of small, screen-ready Web sites provide a multitude of outstanding educational resources, never before used in a portable format.

Once students are used to reading text on a mobile device, the next step follows in Chapter 6, "Writing eBooks and Notes." Not only can students prepare eBooks, but also teachers can prepare personalized assignments and enrichment sources in the form of eBooks.

Noting the potential of eBooks, many libraries are now circulating eBooks. Chapter 7, "Circulating eBooks and eAudiobooks," covers the distribution applications and documents available for download, the advantages of circulating digital media from a school library, locating eBook managing sources, and managing an eAudio library. This chapter provides ideas to assist librarians in establishing a school eLibrary.

Chapter 8, "Utilizing and Preparing eAudio," provides a background as to the audiobook industry, the educational value of audiobooks, and locating audiobooks. The possibilities of converting text to audio for second language learners and the disabled are discussed in depth.

Podcasting has mushroomed in recent months. Chapter 9, "Utilizing and Preparing Podcasts," provides the history, educational uses, legal issues, and the hardware and software needed to locate and listen to a podcast. Guidelines are provided for recording a podcast, along with tips on hosting and publishing that podcast. Legal issues and pitfalls in this fast growing field are covered along with links for further study.

As soon as audio podcasting began to be used, video podcasting was close behind. Chapter 10, "Locating and Downloading eVideos and Vodcasts," covers the brief history of vodcasts and sources for educational eVideos and vodcasts. Selecting software and media plays for video playback is covered along with how to convert video formats for playback on a handheld device and how to prepare video for the micro screen.

Any new development in technology brings along a quagmire of legal issues. Chapter 11, "Digital Media Copyright Issues," addresses some of the issues that library media specialists and teachers confront nearly on a daily basis. The balance between ownership of intellectual property and the fair use of scholarship continues to be a constant challenge. The law struggles to keep pace with the explosion of new technology and the form of the transmission of creative media.

Chapter 12, "Incorporating eBooks, eAudio, eVideo, and Podcasts into the Curriculum," provides ideas for teachers and media specialists to consider in applying the use of mobile media into their school's curriculum. Examples are provided in each of the major curriculum areas as to applying these new forms of media to specific lessons.

Chapter 13, "Record Keeping with Handheld Devices," covers the software available to help media specialists and teachers organize their classrooms and libraries. Access to information anytime and anywhere has streamlined the routine tasks and freed educators to use that time for direct student instruction and class preparation.

chapter one

Using Handheld Devices to Meet National and State Standards

State and national standards for student learning emphasize the importance of a student-centered approach to instruction. Handheld devices offer the promise of engaging students and providing immediate input to the teacher. While the use of handheld computers in schools is hardly widespread at this point, interest is quickly growing as administrators, educators, students, and parents explore the multitude of uses of the handheld computer and how these devices can assist schools, as well as individual students, meet educational standards and informational literacy standards.

In recent months, the use of handheld computers has grown beyond the personal digital assistant (PDA) of a calendar, address book, tasks, and notes, to mobile devices that are also used to read eBooks, listen to eAudiobooks, podcasts, or view eVideos or pictures. While some may fear these uses of handheld computers are strictly for entertainment, wise educators will soon realize the multitude of possibilities for the use of handheld devices within the school.

Two major factors set handheld computers apart from desktop machines: price and portability. Handhelds' slimmed-down software applications mean a quick turnaround time for enhancements and updates. Students and teachers can obtain timely versions of reference resources, recently published books, along with the ability to download daily news and journal articles, beam these resources to others, and read, watch, or listen to the resource without a network connection.

Standards for the English Language Arts

Some people fear that the use of technology may reduce or discourage the use of print materials. However, in many situations it has just the opposite effect. Many educators agree that educational technology was designed to enhance the use of print materials by providing multiple formats to appeal to varied learning styles. The National Council of Teachers of English (NCTE) and International Reading Association (IRA) recognized this when they established the Standards for the English Language Arts.

> "The vision guiding these standards is that all students must have the opportunities and resources to develop the language skills they need to pursue life's goals and to participate fully as informed productive members of society."
>
> <www.ncte.org/about/over/standards/110846.htm>

Figure 1:1 Vision Guide

Using handheld devices within the curriculum will meet the following Standards for the English Language Arts:

"1. Students read a wide range of print and non-print texts to build an understanding of texts, of themselves, and of the cultures of the United States and the world; to acquire new information; to respond to the needs and demands of society and the workplace; and for personal fulfillment. Among these texts are fiction and nonfiction, classic and contemporary works."

"4. Students adjust their use of spoken, written, and visual language (e.g., conventions, style, vocabulary) to communicate effectively with a variety of audiences and for different purposes."

"5. Students employ a wide range of strategies as they write and use different writing process elements appropriately to communicate with different audiences for a variety of purposes."

"8. Students use a variety of technological and information resources (e.g., libraries, databases, computer networks, video) to gather and synthesize information and to create and communicate knowledge."

"12. Students use spoken, written, and visual language to accomplish their own purposes (e.g., for learning, enjoyment, persuasion, and the exchange of information)."

Standards for the English Language Arts, by the International Reading Association and the National Council of Teachers of English, Copyright 1996 by the International Reading Association and the National Council of Teachers of English. Reprinted with permission.

The National Education Technology Plan

The National Education Technology Plan <http://www.ed.gov/about/offices/list/os/technology/plan/2004/site/edlite-background.html> is designed to help motivate and incite a technology-driven transformation in our schools and provide a set of action steps and recommendations that the nation's school systems can consider as they move forward in using technology to meet educational goals and standards.

> "2. Consider Innovative Budgeting
>
> Needed technology often can be funded successfully through innovative restructuring and reallocation of existing budgets to realize efficiencies and cost savings. The new focus begins with the educational objective and evaluates funding requests – for technology or other programs – in terms of how they support student learning. Today, every program in *No Child Left Behind* is an opportunity for technology funding – but the focus is on how the funding will help attain specific educational goals."

Figure 1:2 Innovative Budgeting

The second major action step on page 42 of The National Education Technology Plan states: A handheld for each student can be a cost effective alternative for limited technology funding as the price of handheld computers ranges from approximately $100 per unit up to more than $400, while a single desktop computer costs around $1,000. For little more than the cost of a single textbook, a student can have access to a library of books, in either print or audio format, combined with a personal information manager, and a host of educational software.

The sixth major action plan of The National Education Technology Plan states on page 43:

> **"6. Move Toward Digital Content**
>
> A perennial problem for schools, teachers, and students is that textbooks are increasingly expensive, quickly outdated, and physically cumbersome. A move away from reliance on textbooks to the use of multimedia or online information (digital content) offers many advantages, including cost savings, increased efficiency, improved accessibility, and enhancing learning opportunities in a format that engages today's web-savvy students."

Figure 1:3 Digital Content

Throughout this book, we will discuss methods and techniques essential in moving a school toward the productive use of digital content within the curriculum. Thousands of resources are available for handheld devices, but it is critical to find the exact resource to meet a specific curricular need to help students meet the national and state learning standards.

The National Educational Technology Standards for Students (NETS-S)

The National Educational Technology Standards (NETS) Project is an ongoing initiative of the International Society for Technology in Education (ISTE) and a consortium of distinguished partners and co-sponsors. The primary goal of the ISTE NETS Project is to enable stakeholders in PreK-12 education to develop national standards for educational uses of technology that facilitate school improvement in the United States.

The NETS for Students is the Technology Foundation Standards for all students (NETS-S). By expanding the use of handheld devices to include digital content and format, the following NET Standards will be met:

STANDARD 2: SOCIAL, ETHICAL, AND HUMAN ISSUES

- "Students practice responsible use of technology systems, information, and software.
- Students develop positive attitudes toward technology uses that support lifelong learning, collaboration, personal pursuits, and productivity."

The use of a personal mobile device provides excitement and enthusiasm for students as they learn to take responsibility for its care and maintenance. As students expand their use of handheld devices to include eBooks, eAudio, eVideo, and podcasts, they increase their collaboration tools, along with tools to pursue their independent interests.

STANDARD 3: TECHNOLOGY PRODUCTIVITY TOOLS

- "Students use technology tools to enhance learning, increase productivity, and promote creativity."

Through being able to take full advantage of the multimedia features of a handheld computer, students will be able to increase their productivity and creativity because information obtained through their handheld computer appeals to their multiple intelligences.

STANDARD 4: TECHNOLOGY COMMUNICATIONS TOOLS

- "Students use a variety of media and formats to communicate information and ideas effectively to multiple audiences."

STANDARD 5: TECHNOLOGY RESEARCH TOOLS

- "Students use technology to locate, evaluate, and collect information from a variety of sources.
- Students evaluate and select new information resources and technological innovations based on the appropriateness for specific tasks."

> Reprinted with permission from *National Educational Technology Standards for Students: Connecting Curriculum and Technology,* copyright © 2000, ISTE ® (International Society for Technology in Education), <www.iste.org, iste@iste.org>. All rights reserved. Permission does not constitute an endorsement by ISTE.

Handheld computers can be used as a personal research tool in print, audio, and video formats. Complete encyclopedias, dictionaries, and other reference works can be loaded on a handheld in eBook format. One of the advantages of using eBooks as a research tool is that the information is easily updated and the information is then more current than the same encyclopedia or dictionary in print format on the library shelf.

The National Educational Technology Standards for Teachers (NETS-T)

Not only does the use of handheld computers help meet the Educational Technology Standards and Performance Indicators for Students, but it also helps teachers meet their technology standards, as well.

Standard One of the Educational Technology Standards and Performance Indicators for All Teachers (NETS-T) states:

> "I. TECHNOLOGY OPERATIONS AND CONCEPTS.
> Teachers demonstrate a sound understanding of technology operations and concepts. Teachers:
>> B. demonstrate continual growth in technology knowledge and skills to stay abreast of current and emerging technologies."

For a number of years, handheld computers have been available as personal digital assistants with calendars, memo pads, and address book. However, today educators can use handheld devices to read and research eBooks, listen to audio and podcasts, and watch video on educational topics.

Standard Two of the NETS-T states:
> "II. PLANNING AND DESIGNING LEARNING ENVIRONMENTS AND EXPERIENCES.
> Teachers plan and design effective learning environments and experiences supported by technology. Teachers:
>> A. design developmentally appropriate learning opportunities that apply technology-enhanced instructional strategies to support the diverse needs of learners.
>> C. identify and locate technology resources and evaluate them for accuracy and suitability.
>> D. plan for the management of technology resources within the context of learning activities.
>> E. plan strategies to manage student learning in a technology-enhanced environment."

As teachers locate and evaluate resources for their handheld device, design appropriate learning opportunities, and manage the use of classroom sets of handheld computers, Standard 2 of the NETS-T Standards will be met.

Standard Three of the NETS-T states:
> "III. TEACHING, LEARNING, AND THE CURRICULUM.
> Teachers implement curriculum plans that include methods and strategies for applying technology to maximize student learning. Teachers:
>> A. facilitate technology-enhanced experiences that address content standards and student technology standards.

B. use technology to support learner-centered strategies that address the diverse needs of students.
C. apply technology to develop students' higher order skills and creativity.
D. manage student learning activities in a technology-enhanced environment."

With a little creativity, handheld computers and other mobile devices can be easily incorporated into a technology-rich curriculum. Administrators, teachers, librarians, and students must work together to support learner-centered strategies that address the varied needs of students and require higher order thinking skills. Managing and organizing a classroom set of handheld computers helps teachers meet standard three of the NETS-T as they implement curriculum plans that include methods and strategies for applying technology to student learning that enhances higher order thinking skills.

Standard Five of the NETS-T states:
"V. PRODUCTIVITY AND PROFESSIONAL PRACTICE.
Teachers use technology to enhance their productivity and professional practice. Teachers:
B. continually evaluate and reflect on professional practice to make informed decisions regarding the use of technology in support of student learning.
C. apply technology to increase productivity."

Reprinted with permission from *National Educational Technology Standards for Teachers: Preparing Teachers to Use Technology,* copyright © 2002, ISTE ® (International Society for Technology in Education), <www.iste.org>, <iste@iste.org>. All rights reserved. Permission does not constitute an endorsement by ISTE.

Using handheld computers can increase teacher productivity by providing a ready resource of professional development materials in digital format. Educators can keep their handheld in their purse or clipped to their belt and refer to it whenever they have a few isolated minutes that otherwise would be wasted time.

The Educational Technology Standards for Administrators

It is difficult for an individual teacher or a school to meet technology standards without the enthusiastic support of the administration. Therefore, the International Society for Technology in Education developed the Educational Technology Standards and Performance Indicators for Administrators <http://cnets.iste.org/administrators/a_stands.html>. As school administrators provide leadership for the financing of handheld computers for both faculty and student use, plan curriculum that utilizes the multiple features of handheld computers, and provide professional development for the teachers, the following NET Standards for Administrators will be met.

"I. LEADERSHIP AND VISION.
Educational leaders inspire a shared vision for comprehensive integration of technology and foster an environment and culture conducive to the realization of that vision. Educational leaders:
C. Foster and nurture a culture of responsible risk-taking and advocate policies promoting continuous innovation with technology."

Innovative habits of incorporating handheld devices into the curriculum are beginning to come to the forefront. It takes strong leadership to arrange financial support, professional development, and curriculum changes for a technology that is unfamiliar to many of the stakeholders. However, with research and confident leadership, educators and students will have an opportunity to attempt newer technologies without fear of failure.

"II. LEARNING AND TEACHING.
Educational leaders ensure that curricular design, instructional strategies, and learning environments integrate appropriate technologies to maximize learning and teaching. Educational leaders:
A. identify, use, evaluate, and promote appropriate technologies to enhance and support instruction and standards-based curriculum leading to high levels of student achievement.
C. provide for learner-centered environments that use technology to meet the individual and diverse needs of learners."

In order to assist teachers in enhancing standards-based curriculum, administrators must take the lead in identifying and evaluating the multitude of hardware, software, eBook, eAudio, eVideo, and podcasts available for handheld computers. Administrators can help organize and maintain an environment where students can synchronize their handheld devices with desktop computers in order to add suitable eBooks and eAudiobooks on portable devices with minimal time and frustration.

At the same time, the classroom teacher and library media specialist need to be provided with adequate funding to monitor handheld computer resources and provide selections that would best meet the individual student's needs.

"III. PRODUCTIVITY AND PROFESSIONAL PRACTICE.
Educational leaders apply technology to enhance their professional practice and to increase their own productivity and that of others. Educational leaders:
A. model the routine, intentional, and effective use of technology.
E. maintain awareness of emerging technologies and their potential uses in education."

To be most effective, administrators need to be visible in their expanded use of handheld computers for their personal and professional use and model their use for students. Besides being aware of how other schools are incorporating eBooks, eAudio, eVideos, and podcasts into their curriculum, innovative administrators need to examine specific software and question how it might improve a standards-based curriculum.

"IV. SUPPORT, MANAGEMENT, AND OPERATIONS.
Educational leaders ensure the integration of technology to support productive systems for learning and administration. Educational leaders:
B. implement and use integrated technology-based management and operations systems.
C. allocate financial and human resources to ensure complete and sustained implementation of the technology plan."

Administrators can meet Standard Four of the NET Standards for Administrators by providing finances and training to implement the use of handheld computers into the technology plan. Once handheld devices have been purchased and the teachers and librarians trained, the administrators need to monitor the program as to its effectiveness in meeting technology standards in the curriculum and provide leadership in sharing new software and application possibilities. A well-trained media specialist can be of great assistance to an administrator in sustaining and implementing the technology plan.

"VI. SOCIAL, LEGAL, AND ETHICAL ISSUES.
Educational leaders understand the social, legal, and ethical issues related to technology and model responsible decision-making related to these issues. Educational leaders:
A. ensure equity of access to technology resources that enable and empower all learners and educators."

The lower cost of handheld devices makes it easier for cash-strapped schools to provide equal access to handheld computers for both students and teachers than to provide laptop or desktop computers.

Monitoring copyright issues pertaining to resources in the digital format can be challenging for both administrators and teachers, but as school administrators rise to the challenge, they meet Standard Six of the NET Standards for Administrators.

Reprinted with permission from *National Educational Technology Standards for Administrators,* copyright © 2002, ISTE ® (International Society for Technology in Education), <www.iste.org>, <iste@iste.org>. All rights reserved. Permission does not constitute an endorsement by ISTE.

Information Literacy Standards

The use of handheld devices in a school environment lends itself to meeting Information Literacy Standards as set forth by the American Association of School Librarians and Association for Educational Communications and Technology in a broader, more specific manner than many of the other technology resources available to schools at this time.

"STANDARD 1: THE STUDENT WHO IS INFORMATION LITERATE ACCESSESS INFORMATION EFFICIENTLY AND EFFECTIVELY."

> **Indicator 4. Identifies a variety of potential sources of information**

Figure 1:4 Indicator 4 Sources

A large assortment of eBooks, eAudio, eVideo, and podcasts are available for mobile devices. Understanding the criteria in evaluating and selecting these resources plus knowing where to locate those resources, is critical for information literacy and effectiveness.

"STANDARD 3: THE STUDENT WHO IS INFORMATION LITERATE USES INFORMATION ACCURATELY AND CREATIVELY."

> Indicator 4. Produces and communicates information and ideas in appropriate formats.

Figure 1:5 Indicator 4 Information

Several digital formats are available for handheld devices to assist users in communicating ideas and information. With proper instruction and creativity, students can expand and share their knowledge in a variety of subject areas.

"STANDARD 4: THE STUDENT WHO IS AN INDEPENDENT LEARNER IS INFORMATION LITERATE AND PURSUES INFORMATION RELATED TO PERSONAL INTERESTS."

> Indicator 1. Seeks information related to various dimensions of personal well-being, such as career interests, community involvement, health matters, and recreational pursuits
>
> Indicator 2. Designs, develops, and evaluates information products and solutions related to personal interests

Figure 1:6 Indicators 1 and 2

One of the greatest advantages of every student having access to their own handheld computer is the ability to become an independent learner. If able to select and access eBooks, eAudio, eVideo, and podcasts from a library or school Internet site, students will be motivated to do personal research in their areas of interest and need.

"STANDARD 5: THE STUDENT WHO IS AN INDEPENDENT LEARNER IS INFORMATION LITERATE AND APPRECIATES LITERATURE AND OTHER CREATIVE EXPRESSIONS OF INFORMATION."

> Indicator 2. Derives meaning from information presented creatively in a variety of formats
>
> Indicator 3. Develops creative products in a variety of formats

Figure 1:7 Indicators 2 and 3

The use of handheld computers increases the appreciation of literature, as students are able to read an eBook or listen to an eAudiobook in locations and time allotments that might not be possible with a traditional print book. Libraries can be the site for downloading recommended, quality eBooks.

Summary and Challenge

If utilized properly, handheld devices have the capabilities to meet a large majority of curriculum and technology standards to help students raise their standardized test scores. A handheld device can contain a library "in the palm of a student's hand" to serve as a research and a ready-reference tool. With these devices, content knowledge and skills related to curriculum, instruction, and assessment can receive high priorities as compared to the limitations of print material only, accompanied with limited use of desktop computers.

Not only can students' test scores be improved with the use of handheld devices, but also teachers, media specialists, and administrators can improve their teaching quality standards, while they better utilize their time and funding considerations. Every educator needs to consider the following questions: How does your school compare with the various national standards? Will increasing the use of handheld devices help your school better meet those standards? Locate a copy of your state and local technology standards and describe how the use of handheld computers might help you meet those standards. After you finish this book, return to your list and make modifications as to how handheld technology will enhance your technology-rich curriculum.

chapter two

Selecting Hardware for eBooks, eAudio, eVideo, and Podcasting

Students often feel a disconnection between their personal use of technology and their technology experiences in school. The desktop computer is seen as "their father's computer" while their computer is a condensed, multimedia, multifunction device. The children in our schools today are becoming known as the "thumb generation" as they spend more time entering data into a handheld device as opposed to a desktop computer. With limited funding, schools must search for ways to overcome this technology disconnect and apply current entertainment devices to educational purposes.

Handheld computers have come a long way from the older 160 x 160 pixel black-and-white or gray scale screens to 640 x 480 (or greater) color screens. Handhelds can now do 80 percent of what a laptop can do at 10 percent of the cost.

Types of Handheld Devices

In selecting handheld computer devices, one needs first to consider which operating system will best meet the needs of students, teachers, and administrators. There are two major operating systems in handheld computers: the Palm® OS from Palm, Inc. and Windows Mobile® from Microsoft. Windows Mobile is a watered-down version of Microsoft Windows® for desktop computers. The Windows CE operating system was a forerunner of Windows Mobile. Windows Mobile 2003 Second Edition operating system is the most common, but newer models are being released with a Windows Mobile 5.0 operating system.

The Windows Mobile software platform includes more platform flexibility to customize devices and solutions. This operating system contains productivity enhancements that include: updated Microsoft® Office software, persistent memory storage for more efficient data management, a powerful multimedia experience with Microsoft's Windows Media Player 10, and support for hard drives.

The other operating system is the Palm OS. Until recently, handheld computers compatible with the Palm OS operating system were simple, easy-to-use devices, but were weak in areas like multimedia when compared to handhelds powered by Windows Mobile. However, with the release of Palm OS Garnet 5.4, many multimedia features were added.

In comparing the operating systems, Windows Mobile has more versatility, can contain built-in wireless, and is generally more cost effective. It is more like a traditional desktop computer. However, the Palm OS is a better operating system for lower elementary grades, because it is less complicated and is an easier-to-manipulate operating system. The high-end Palm handhelds can also contain wireless connectivity.

Although reviews say that neither Palm OS nor Windows Mobile has a huge advantage, one factor might still tip the scales. Windows Mobile is still not compatible with Macintosh® computers, so users will not be able to use the Apple® iSync™ application to synchronize a Windows Mobile handheld with a Macintosh computer. However, third-party software, such as Chapura's PocketMirror®, can be purchased for around $75, and it will let users synchronize a handheld using Windows Mobile software with a Macintosh computer. Right now, without supplemental software, only a Palm OS handheld computer will work with a Macintosh computer.

First-time handheld computer users may feel overwhelmed with the wide variety on the market. A concise comparison of these devices can be found at consumersearch.com <www.consumersearch.com/www/computers/pda-reviews/comparisonchart.html> or PC World <www.pcworld.com/resource/browse/0,cat,1167,sortIdx,1,pg,1,00.asp>.

Some eBooks, eAudio, and eVideo are available for smartphones even though the ability to play audio or video is limited and the smaller screen makes reading an eBook difficult. A smartphone is either a cell phone with handheld computer capabilities or a traditional handheld computer with added cell phone capabilities, depending on the style and manufacturer. Smartphones are becoming increasingly popular; however, it can be difficult to select a smartphone because not every phone works on every wireless cell phone network.

Cellular service providers handle phone service for smartphones. As with cell phones, consumers typically purchase both a cellular plan and smartphone from the same service provider. A number of operating systems—Windows Mobile Pocket PC Phone Edition, the Palm OS, the BlackBerry® OS for BlackBerry smartphones, and the Symbian® OS for smartphones from Panasonic, Nokia, Samsung, and others—may contain features that can be programmed to allow cell phone connections.

Motorola has introduced the ROKR®, the first iTunes® phone that stands out with its seamless compatibility with iTunes. The ROKR will allow users to pause their audio when an incoming call comes and then resume playing audio where it left off. The ROKR supports MP3 and AAC formats used on Macintosh computers, as well as protected AAC audio formats and video clip capture and playback. As with other combination units, the smaller screen and limited storage space limits the multimedia use of the ROKR.

The Apple iPod and the iPod nano® are becoming favorites with the net generation. Originally, the iPod was designed as a jukebox to play MP3 format music, but eAudiobooks, eVideos, and eBooks are now available for the iPod. The Apple Corporation is now working on agreements with major studios and record labels over copyright protection and anti-piracy technology to provide movie videos for the iPod.

The iPod nano is 80 percent smaller than Apple's original iPod and 62 percent smaller than the iPod mini. Apple's iPod nano is a flash-based player, which weighs 1.5 ounces and measures about a quarter-inch thick. The iPod nano's screen is only slightly larger than a postage stamp, and on many MP3 players that would mean users could not see the entire name of many of the albums or tracks they play. However, the iPod nano's screen resolution is high enough that it can fit as many as 27 characters across the screen in type that is extremely readable.

The other remarkable aspect of the iPod nano is its capacity. Most flash-based players top out at 1GB, but the iPod nano is in the mid-capacity range that, until now, has been used exclusively by portable media players with internal hard drives. Some users can hold more than 400 songs (encoded at anywhere from 128 kbps to 192 kbps) on the 2 GB model of the iPod nano. Using flash memory makes the iPod nano more rugged and appealing to joggers, bicyclists, and others who may have been reluctant to risk ruining a hard-drive player during their exertions. The 4 GB version of the iPod nano sells for around $250 while the 2 GB version is $199.

Sony's PlayStation® Portable <www.us.playstation.com/> is not only a gaming device, but also provides videos for download in both the QuickTime® and the Windows Media® format.

Selection Consideration Factors

When selecting a handheld device, besides the operating system, the following factors should be taken into consideration:

Desired Features

Will the users be interested in strictly an eBook Reader, an eAudio player, or an eVideo player, or will they also need a personal information manager (PIM) that includes calendar, contacts, email, and tasks? Will the handheld device also need to double as a global positioning system (GPS), a camera, or the source of power for scientific probes or other educational resources?

Price

Handhelds utilizing the Palm OS range from $99 to $600. Handhelds with the Windows Mobile platform begin at $250.

Battery Life

In a school environment, teachers and administrators will want handheld devices that have longer-lasting, rechargeable batteries. Palm handhelds used to run on two AAA batteries, but the demands of today's fast processors and color displays are too intensive for AAAs to handle.

Lithium Ion batteries vary in capacity, but most standard batteries contain around 1,000 mAh (mAh stands for Milli-Amp Hour. It is a measure of the capacity, or "amount of power," a battery can hold.). These batteries last an average of two years before replacement is necessary due to significantly diminished run times.

Palm OS handhelds with 200 MHz or slower processors will last a few days of average use (about eight hours actual use) per charge. Pocket PCs running Windows Mobile software last around three hours of actual use per charge. These runtimes will vary, depending on what the user does with the handheld; games and multimedia are much more demanding than a personal information manager, Office applications, and eBooks. Wireless networking such as Wi-Fi® and Bluetooth® also consume power and can cut battery life by 40 percent.

Some handhelds can be damaged because of infrequent charging of the battery, which could lead to the battery dying and/or data loss. Whenever possible, avoid going several days without charging a handheld. It is preferable to keep a handheld computer charging in its cradle.

Display

Screen sizes generally range from 2.25-inch by 3.125-inch to the 2.25-inch by 2.25-inch display, some up to four inch square. There are two screen types available: color and two-tone grayscale. Those who plan to use a handheld computer for PowerPoint® presentations, Web browsing, or viewing movies and pictures may want to consider a color screen. These generally cost about $100 more than the grayscale versions.

Size and Weight

Generally, Pocket PCs are larger and heavier than Palm handhelds but that is beginning to change as newer Pocket PCs become smaller. PDA sizes and weights are constantly changing. Currently, most handhelds weigh between three and seven ounces, and their height and width can range from four to six inches and two to four inches, respectively. The size of a handheld can increase significantly with the addition of hardware such as a modem or wireless card, particularly if the handheld requires a specialized expansion pack.

Handwriting Recognition

Both Palm OS and Pocket PC handhelds contain handwriting recognition and on-screen keyboards. Palm OS handhelds use Graffiti® handwriting recognition while devices running Palm OS 5.2 or newer feature Graffiti 2, which uses a natural print alphabet. Pocket PCs offer three handwriting recognition options: Block Recognizer, which is the same as Graffiti (great for long-time Palm users who switch to Pocket PC); Character Recognizer, which uses natural alphabet printing similar to Jot® and Graffiti 2; and Transcriber, which allows users to write in cursive/script. Overall, Transcriber is not as accurate as Jot and Graffiti 2.

Input/Out(I/O) Ports

The handheld platform dictates that all handheld devices have several input/output ports to provide the maximum flexibility to communicate with the outside world. Examples of I/O ports include USB ports to move data from desktop to handheld. All handheld devices include an infrared port that can be used to communicate with other handheld devices via an invisible beam of light. This feature is an invaluable tool for students to collaborate with others or their teachers without printing the text each time it is changed.

Multimedia Capabilities

Pocket PCs had more multimedia features than the Palm, but with the release of Palm OS 5, multimedia is becoming available for the Palm operating system as well. Many handhelds double as MP3 players using a headphone set. The Apple iPod is recognized for its multimedia features.

Voice Recorder

Pocket PCs and some Palm handhelds double as voice recorders. Inputting data on a handheld computer can be a difficult experience with tiny virtual keyboards or inaccurate handwriting-recognition programs. Therefore, a built-in voice recorder can be extremely helpful. With the press of a button, a user can dictate reminders, ideas, or passing thoughts into a handheld device. Some students use their handheld devices to record class lectures.

Microprocessor Speed

The processor is the brain of the handheld device. Currently, Pocket PC processor speeds are between 200 and 400 MHz, while Palm OS processor speeds are between 33 and 66 MHz. These differences can be deceptive. The Palm OS is highly optimized and requires less power, while the Pocket PC, based on Microsoft Windows, requires more resources. As of this writing, the processor speeds on handheld computers ranged as high as 624 MHz.

Memory and Expandability

Handheld computers do not have hard drives, but store all files and information in their memory. Pocket PC devices range from between 32 MB and 64 MB of built-in memory expandable to 1GB via expansion cards while Palm devices range from 2 MB to 16 MB of built-in memory, expandable to 1GB via expansion cards. Because of these constraints, handheld computers are not able to hold as many large files such as MP3s or video files without an expansion card. As more programs, eBooks, eAudio, and eVideo are added to a handheld device, less memory is available and the handheld's performance decreases. The Apple iPod does contain a hard drive with up to 80 GB of storage space.

Expansion Slots

Some handheld devices have expansion slots that can accept certain types of cards from SanDisk Corporation such as the CompactFlash® (CF), usually the least expensive card; the Reduced Size

MultiMediaCard™ (RS-MMC), the smallest card in size; SecureDigital™ (SD), the fastest card; or Sony's Memory Stick™.

Figure 2:1 CompactFlash Card

Figure 2:3 Secure Digital Card

Figure 2:2 MultiMediaCard

Figure 2:4 Sony Memory Stick

Depending on the type, these slots may be used for additional storage, or to attach peripherals such as modems, wired and wireless network adapters, or global positioning system (GPS) devices. Some devices have no slots built into the body of the device, but expansion jackets that include them are available as accessories. Users can store applications on flash memory cards, but applications run slower on flash memory cards than when stored in internal memory.

Backup

All handheld devices have at least one of two types of interfaces for making a physical connection with a host PC to synchronize data and upload programs to desktop or laptop computers: serial or Universal Serial Bus (USB). Serial connection is an older, slower standard. USB is more recent and supports faster transfer of files and applications, and it is now much more common than serial connections. USB2 connectors are now becoming popular with the most recent handheld devices.

Some handheld devices include a docking cradle into which the device can be inserted to synchronize data and download programs from the host PC and/or recharge the batteries. The cradle remains connected to the host PC when the device is removed. Models that are more basic include only a data cable, not a full cradle.

Many handheld devices include an infrared port (often called an IRDA port named for the InfraRed Data Association) that can also be used to synchronize data with a host PC or to exchange data with other devices. Devices must be within a few feet of each other with a clear line of sight between them for beaming from infrared ports to work.

A number of handhelds (including both Pocket PC and Palm OS models) now include built-in wireless connectivity or Bluetooth technology to connect with printers and other such devices.

Peripherals

Both the Palm and Pocket PC are compatible with a number of devices, including scientific probes, digital cameras, global positioning systems, network cards, phones, MP3 players, and wireless communication. While expansion items may seem like an unnecessary luxury when one first begins to consider the use of handheld devices within the curriculum, if purchases are made with consideration of possible expansion, schools will get a much greater value from their investment than if they purchase for only their immediate needs.

Keyboards

Both Palm OS and Pocket PC devices have portable keyboards for easier data entry. Portable keyboards are desirable if the user plans to use his handheld computer to compose email, edit or compose Word documents, or enter large amounts of data. Full-size, portable keyboards are available that closely resemble a standard keyboard. Some of these are wireless and use the PDA's infrared port to connect. There are also smaller keyboards available that work well, but their reduced size can take time to become accustomed to using.

Figure 2:5 External Keyboard

Several brands of virtual keyboards for handheld computers are beginning to appear on the market.

Figure 2:6 Virtual Keyboard

Modems and Connectivity

Both Palm and Pocket PC handhelds can use modems to connect to the Internet if the appropriate slots are available. Wired Ethernet or Wi-Fi using CompactFlash (CF) cards or SecureDigital (SD) cards can be available if the appropriate slots are present on the handheld. Some devices contain either built-in Bluetooth connectivity or they can be added with an appropriate SD card. There are no Bluetooth SD cards for Palm OS 5 handhelds.

Two forms of wireless connectivity that are particularly relevant to handhelds are Bluetooth and 802.11 wireless LAN. Some wireless connections also double as a cell phone. Wi-Fi lets the user pick up email and surf the Internet at hotspots in coffee shops and airports; and General Packet Radio

Service (GPRS) lets a handheld connect to the Internet from anywhere for a monthly fee. GPRS is a mobile data service available to users of Global System for Mobile communications (GSM) mobile phones.

Bluetooth communication is used primarily as a wireless replacement for a cable to connect a handheld, mobile phone, MP3 player, printer, keyboard, mouse, or digital camera to a PC or to each other, while IEEE 802.11 supports greater distances and is typically used to connect devices to a building's wireless LAN.

GPS Systems

To take advantage of the power of a handheld computer, Global Positioning Systems (GPS) are available for handheld computers with CompactFlash expansion cards. Some of the more popular ones include Mapopolis: Maps for Handhelds <www.mapopolis.com/>, Garmin iQue Series GPS <www.tigergps.com/pdacompanion.html>, Pocket CoPilot GPS <www.alk.com/copilot/>, and Destinator GPS <www.destinatortechnologies.com/>.

Scientific Probes

Scientific probes offer an inexpensive way to include a variety of technology in a classroom for use with hands-on laboratory activities. Students learn how to use scientific probes in the classroom with handheld computers. Laboratory experiments can be available using light sensors, temperature probes, force plates, drop counters, CO_2 gas sensors, motion detectors, radiation monitors, gas pressure sensors, and EKG sensors.

Sensors can be important to the curriculum because they provide highly interactive learning experiences without the drudgery often associated with labs; students can immediately see multiple representations of data while an experiment is underway.

ProbeSite <http://probesight.concord.org/> provides an outstanding online resource for curriculum templates and suppliers of scientific probes for the K-12 environment.

Summary and Challenge

With the ever-increasing number of handheld devices on the market and with each model containing its own unique features, selecting the best handheld device can sometimes feel overwhelming. Generally, the more features, the higher the cost. However, a handheld device with more features and the ability to add more peripherals later can often be the best investment overall. Memory is the first consideration in the selection of a handheld device and corners should not be cut in its selection.

Before purchasing handheld devices for your school, or personal use, review the options available. The questions to consider are: "What operating system will best meet the particular needs of the user?" "What features on an individual device will be best for the particular situation?" "Which handheld computer accessories do you anticipate using within the next 24 months?" and "Will funding have an impact on your selection? If so, what are some alternative sources of funding?"

chapter three

Selecting Software for the Handheld Device

Handheld computers generally come pre-installed with several software packages along with a CD containing additional software. Before purchasing a handheld computer, be sure to verify whether it comes with the types of software that will be needed, or if the software is available to be purchased separately.

Synchronizing Handheld Computers to Desktop or Laptop Computers

The process of synchronizing data onto a handheld computer is very simple. On a Palm handheld, after the data are loaded onto a desktop or laptop computer, the user merely places the handheld computer in its cradle and presses the HotSync® button located on the cradle of the handheld device. This simple procedure will begin the synchronization process and move the data from the personal computer onto the handheld computer.

To synchronize files from the desktop using Microsoft Windows to a Pocket PC, Microsoft ActiveSync® <http://www.microsoft.com/windowsmobile/downloads/activesync42.mspx> must be loaded on the desktop computer. To synchronize files from the desktop using Microsoft Windows or from a Macintosh to a Palm Operating System, HotSync software is necessary <http://www.palm.com/us/support/software.html>.

It is possible to synchronize digital media files to locations other than a storage card. For best results, users should insert a removable storage card, such as a SanDisk's SecureDigital (SD) or miniSD™ card, into the device and then synchronize to that location. Synchronizing to another storage location, such as to the device's RAM, can be problematic for certain devices.

Document Preparation Software

Word processing software is often considered the basis of digital print communication, while spreadsheet software is the basis for numeric digital manipulation. Handheld computers can utilize streamlined versions of popular word processing and spreadsheet software for desktop computers.

Pocket PCs integrate Microsoft Word and Excel® documents into its system and can read and write existing desktop files. Users can copy files to their Pocket PC, and back to their desktop as needed, without the advanced features that are part of desktop Office.

The Palm operating system does not have built-in Word and Excel support. To synchronize a Palm operating system handheld with the Microsoft Office suite of Word, Excel, and PowerPoint, a third party software such as Documents To Go®, from <www.dataviz.com/>, is necessary. Documents To Go allows users to edit Word, Excel, and other files directly onto a handheld computer using Chapura's PocketMirror software <www.chapura.com>.

Recently other personal information management software has appeared on the market to compete with the Microsoft Suite. Pocket Informant® <www.pocketinformant.com/> replaces the built in Contacts / Calendar / Tasks applications with a more powerful, yet simple, full-featured version along with cool additions in one fully integrated application.

Printing from Handheld Devices

Printing and transferring documents is one of the most useable options of a handheld device. Bachmann Software is a leader of wireless printing and network file exchange <www.bachmannsoftware.com/education.htm>. PrintBoy from Bachmann Software lets users print documents, spreadsheets, presentations, and email directly to a printer using an infrared, Bluetooth, Wi-Fi, serial, or TCP/IP connection. FilePoint software from Bachmann Software permits wireless access to the user's desktop, LAN, or remote network computers to exchange documents, music, audio, and photos.

Standards for Digital Media

In 1984 the Center for Applied Special Technology, CAST, <www.cast.org/about/index.html> was established through a collaborative agreement with the U.S. Department of Education's Office of Special Programs. Its mission is "to expand learning opportunities for all individuals, especially those with disabilities, through the research and development of innovative, technology-based educational resources and strategies." CAST's interest in innovative technology in education opened the way for software developers to search for ways to improve access to the general curriculum for students with disabilities.

As technology developed through the next decade, in order to prevent the "Wild West" scenario in digital formatting, those with a vested interest in the field set industry standards. The International Digital Publishing Forum (IDPF), formerly the Open eBook Forum (OeBF), the trade and standards association for the digital publishing industry, sets eBook standards. Their members are of academic, trade, and professional publishers; hardware and software companies; digital content retailers; libraries; educational institutions; and accessibility advocates and related organizations whose common goals are to advance the competitiveness and exposure of digital publishing.

The national government provides a number of agencies to help educators incorporate technology into the curriculum to meet the needs of the disabled. The CPB/WGBH National Center for Accessible Media (NCAM) <http://ncam.wgbh.org/> is a research and development facility dedicated to the issues of media and information technology for people with disabilities.

Audio Standards for Digital Talking Books

Digital Talking Books (NISO Z39.86) sets the standard for format and content of digital talk books (DTB) and establishes a limited set of requirements for DTB playback devices. For many years, "talking books" have been available to print-disabled readers in analog media such as phonograph records and audiocassettes. These formats served their users well in providing human-speech recordings in a robust and cost-effective manner. However, analog media are limited in several features when compared to a print book.

- Analog media offer a linear presentation, which is limited in its flexibility in searching reference works, textbooks, magazines, and other materials that generally are accessed randomly. In contrast, readers can move around in a book or magazine in digital media more freely and efficiently than a sighted reader can flip through a print book.

- Analog recordings do not allow users to interact with the book by placing bookmarks or highlighting material to designate locations. A digital talking book offers this capability by storing the bookmarks and highlights separate from the text of the DTB itself.

- Talking book users have long complained that they do not have access to the spelling of the words they hear. Now many Digital Talking Books will include a file containing the full text of the work, synchronized with the audio presentation, thereby allowing readers to locate specific words and hear them spelled.

- Analog audio only provides readers and listeners one version of the document. If an audiobook contains footnotes, they might be read where they are referenced, which burdens the casual reader with unwanted interruptions, or the footnotes might be grouped at a location out of the flow of the text, making them difficult for interested readers to access. A digital talking book allows the user to easily skip over or read footnotes.

Since its beginning in May 1996, the DAISY (Digital Accessible Information SYstem) Consortium led the worldwide transition from analog to Digital Talking Books. The DAISY standard provides flexibility and options never before possible. There are many kinds of DAISY books, ranging from audio only to text only, with full text and full audio as the fullest and richest reading experience for the end-user. It is even possible to produce hard copy Braille from a full-text DAISY book.

Although the DAISY standard was first developed with books for people with print disabilities, other applications to ensure access to information for all students can be applied to those standards. Imagine having lectures that are available as both audio and text files simultaneously. Many handheld computers now have the capability to meet the requirements of the DAISY standard and provide audio and text simultaneously.

The digital talking book, DTB, offers the print-disabled or slow reader a significantly enhanced reading experience—one that is much closer to that of the sighted reader using a print book. The DTB includes not just the audio rendition of the work but the full textual content and images as well. Because the textual content file is synchronized with the audio file, a DTB offers multiple sensory inputs to readers, a great benefit to learning-disabled or dyslexic readers.

Some visually impaired readers may choose to listen to most of the book but find value inspecting the images that provide additional information not available in the narrative flow. Others may choose to skip the audio presentation altogether and instead view the text file via screen-enlarging software. Braille readers may prefer to read some or the entire document via a refreshable Braille display device connected to their DTB player and accessing the textual content file.

Digital Talking Books are not limited to a single distribution medium. CD-ROMs were the first digital format, but today DTBs are more portable and can utilize MP3 format or any other digital format associated with digital audio recordings. Digital Talking Books will normally be in the context of a digital rights management system, which grew from a desire to standardize DTB file structures in the hope that it might prevent a recurrence of the multiple formats currently used for talking books throughout the world.

In recent years, audiobooks on cassette became popular with commuters, so travelers and listeners were able to hear books that they might not have had time to read. With the advent of the CD player, audiobook readers gradually shifted to the digital format. Educators began to incorporate audiobooks into the curriculum for those who were having problems learning to read or for classroom enrichment. Audiobooks have now become an excellent time management tool for both children and adults.

eBook Readers for Handheld Devices

Each eBook-reader software program has its own format. As a result, users need to make sure that their eBooks are compatible with the eBook reader they have. Some plain text, generic formats of eBooks can be read by a variety of eBook readers, such as DOC. (Note: This is not the same as the .doc format for Microsoft Word documents.) One drawback of DOC files is that they do not have special features like links or pictures.

Before using a handheld computer as an electronic book, it is necessary to have the appropriate eBook reader loaded on the handheld. Most eBook readers are free or inexpensive downloads and can be downloaded when a book in that format is desired. The most common eBook readers and their features, along with where they can be obtained are listed in the table below.

Table 3:1 e-Book Comparisons

Name	Acrobat Reader 7	Microsoft Reader	Mobipocket Reader	E-Reader Pro (formerly Palm Digital Media)
Cost	Free	Free	Free (with 14-day free trial of Mobipocket P) Mobipocket Reader Pro - $19.95	Two versions: a free version, and a Pro version that adds support for additional fonts, color themes, reference books, and formatting options $9.95
Operating Systems	Palm OS, Pocket PC, Symbian OS, Windows, Macintosh, Linux Solaris	Windows, Windows Mobile	Windows, Windows CE, Windows Mobile, Palm	Palm, Pocket PC, Windows Mobile, Symbian
Format	.pdf	.lit	.prc	.pdb, .doc
Images/ multimedia support	PNG, JPG, GIF. Embeds QuickTime movies, Real, WMP (Windows/Mac only) PNG, JPG, GIF. Embeds QuickTime movies, Real, WMP (Windows/Mac only)	PNG, JPG, GIF	BMP, GIF, MP3	JPG, GIF, BMP
Source	http://www.adobe.com/products/acrobat/	http://www.microsoft.com/reader/	http://www.mobipocket.com/	http://www.ereader.com/products/ereader/pro/
Color	Yes, depending on the hardware used	Yes, depending on the hardware used	Yes, depending on the hardware used	Yes, depending on the hardware used
Adjustable type size	Yes	Yes	Yes	Yes
Two-page view	Yes	No	No	No
Printable text	Yes, with publisher permission	No	No	No
Fully text searchable	Yes	Yes	Yes	Yes
Ability to take notes and bookmark	Yes, you can take notes, but you cannot bookmark	Yes	Yes	Yes
Ability to highlight	Yes	Yes	Yes	Yes
Photos and art supported	Yes	Yes	Yes	Yes
Ability to draw in e-book	No	Yes	Yes	Yes
Ability to create e-Books	Must use commercial products such as: Acrobat 7.0 Elements, Acrobat 7.0 Standard, Acrobat 7.0 Professional	Author can use other formats, then convert using ReaderWorks Standard or Publishor http://www.readerworks.com). The Read in Microsoft Reader (RMR) add-in for Microsoft Word enables users to convert any Word document into a Microsoft Reader format eBook. (http://www.microsoft.com/reader/downloads/rmr.asp).	With Mobipocket Creator 4.0 Publisher Edition, create ebooks by importing PDF or Word documents, and target customers who read on their PDA, smartphone or PC.	The DropBook utility is a free download. e-Book Studio creates e-Books that can be read by the eReader and eReader Pro software on Palm OS or PocketPC handhelds.
Misc.	E-books may be read aloud via text-to-speech if publisher enables this feature (desktop version only).	Text-to-Speech Component No embedded multimedia. Linked multimedia OK (desktop version only). Speech output identifies images with alt text.		Supports a large number of special features, such as text formatting, hyperlinks, and pictures.

Table 3:1 e-Book Comparisons

Name	Tomeraider	iSilo	Namo HandStory
Cost	Free	30 day free trial - $19.95	Free Trial; Full Version from $29.95; Upgrade Version from $7.95
Operating Systems	Palm, Pocket PC, Windows, P800/ P900, Psion, Nokia Communicator	Palm OS, Pocket PC, and Windows CE and Pocket PC	Palm or Pocket PC
Format	.tr3, Compatible with .doc	.pdb	supports converting text files in the Unicode format to Palm memo or Doc format
Images/ multimedia support	Included	Images can be many multiples the size of the screen in either dimension, with a theoretical size limit of about 32,000 by 32,000 pixels.	You can convert BMP, GIF, and JPEG files with HandStory Converter to and only to 16 gray format.
Source	http://www.tomeraider.com/	http://www.isilo.com/	http://www.namo.com/products/handstory/
Color		iSilo can display 8-bit (256 colors) and 16-bit (65,000+ colors) color images.	Yes, depending on the hardware used
Adjustable type size		Yes – Four font sizes	Yes
Two-page view		No	No
Printable text		Yes. Can disallow this permission to prevent users from printing a document content.	Yes
Fully text searchable		Yes	Yes
Ability to take notes and bookmark		Bookmarks	Yes
Ability to highlight		Individual blocks can have their own background color, effective for visual highlighting of content such as table headings.	Yes
Photos and art supported		Yes	Yes
Ability to draw in e-book		Yes	Yes
Ability to create e-Books		No	Namo HandStory Converter for your PC is a versatile converter of texts, images, and Web content.
Misc.	Also includes database features	*iSilo* supports VFS, the standard interface for accessing file systems on external media, enabling users to view documents stored on memory expansion cards.	For Palm or Pocket PC displays texts, eBooks, Images, and Web Clips, all from including even Video Clips (for Pocket PC only) within a single application. Namo HandStory Converter for your PC is a versatile converter of texts, images, and Web content.

Media Players for Handheld Devices

Windows Media Player

Windows Media Player for Pocket PC from Microsoft supports all the variants of the Windows Media Audio 9 codec. The desktop version of Windows Media Player 9 Series is not as full-featured as the desktop version of Windows Media Player 10. The desktop version of Windows Media Player 9 Series does not support

> MPEG is to video what MP3 is to music.

Figure 3:1 MPEG

automatic synchronization, synchronization of playlists, synchronization of album art, or synchronization of TV shows recorded by computers running Windows XP Media Center Edition. Thus, handheld devices running Windows Media Player 9 are more limited than those with Windows Media Player 10.

There have been several updates to Windows Mobile 2003 Second Edition. Some Windows Mobile 2003 Second Edition devices might not include Windows Media Player 10 Mobile because only the most recent update (known as Windows Mobile 2003 Second Edition Adaptation Kit Update 2) includes Windows Media Player 10 Mobile.

Windows Media Player 10 Mobile is able to play digital audio and video files that have been copied to a removable storage card, or to play digital audio and video files that are streamed over the Internet. By default, users can play files in either Windows Media or MP3 format (this includes files with the extensions .asf, .wma, .wmv, and .mp3). Windows Media Player 10 Mobile is not available for download from Microsoft, but it is up to each device manufacturer to select which version of the operating system (and, by extension, which version of Windows Media Player) they will include on their new devices. It is also up to each device.

Through a process called synchronization, the desktop version of Windows Media Player 10 lets users copy music, videos, recorded TV, playlists, and pictures to Windows Mobile-based devices. The Synchronize feature of a handheld computer detects the playback capabilities of the device and, when possible, converts the file into a format and bit rate appropriate for the device. All synchronized media items, except pictures, will appear in the Windows Media Player 10 Mobile library on the handheld device.

Protected files are digital media files that are secured with a license to prevent unauthorized distribution or playback. The technology used to protect files is called digital rights management (DRM). Windows Media Player 10 Mobile supports Windows Media DRM 10 (previously known by the code name Janus). If the license permits it, Windows Media Player 10 Mobile can play protected .wma files that users acquire from online stores, such as MSN Music or Napster, through a la carte purchase or through a subscription account. To play subscription content on a handheld device, users must have the desktop version of Windows Media Player 10 in order to synchronize the files to the device.

If the file is protected and the license permits synchronization to a portable device, the Sync feature on a handheld device copies the license to the device and stores it in the appropriate location. The license specifies how the file can be used. For example, a license can specify how many times a file can be played or whether the file can be burnt to a CD or synchronized to a portable device. The person or company that provided the file specifies the terms of the license.

In order to play video on a handheld computer, some newer devices have larger displays that support VGA resolutions (480 x 640 pixels). If users try to use Windows Media Player 10 Mobile to play a QVGA (240 x 320 pixels) video in full-screen mode on a VGA device, depending upon the device and the codec that was used to encode the content, the video might not be scaled to fill the entire screen. The video might only be displayed at 240 x 320 pixels.

Windows Media Player 10 Mobile can play files created by Microsoft Photo Story 1, Photo Story 2, and Photo Story 3 <www.microsoft.com/windowsxp/using/digitalphotography/photostory/default.mspx>. This simple program can create slideshows using digital photos, add special effects, soundtracks, or voice narration to photo stories, and then personalize the file with titles and captions. Small file sizes make it easy to use on a Windows Mobile–based portable device. Users can import the Photo Story 3 .wmv file into Windows Movie Maker, and then save the movie by using specific movie settings that produce the highest-quality files for Pocket PC.

RealPlayer

RealPlayer® is available for any Pocket PC device that meets the following minimum system requirements: Intel® StrongARM or Xscale processor (200 MHz or higher), 32 MB ROM, 32 MB RAM, and 240 width x 320 height screen size. Functionality of the RealPlayer for Mobile Devices varies across devices. RealPlayer for Palm runs on the latest Palm-branded OS 5 devices with ARM processors. It supports the local playback of MP3 and RealAudio® content from inserted SD memory cards. A plug-in for the RealPlayer for PC allows easy and seamless transfer of songs to the device.

Flash Player for Pocket PC

Macromedia Flash® is a player for delivering rich streaming content over low-bandwidth connections <www.macromedia.com/software/flashplayer_pocketpc/>. Flash is designed to deliver a rich media environment while minimizing file size, which makes it a very useful development tool for the Pocket PC. Prior to Flash 6, Flash content for the Pocket PC had to be viewed as an HTML page within Pocket Internet Explorer®. To compensate for this limitation, third-party software packages called FlashAssist™ by Ant Mobile Software <www.antmobile.com> and FlashPack™ by HandSmart software <www.handsmart.com> were developed.

The move to the Flash 6 player offered developers the options available in the Flash MX authoring environment. One of the most important usability enhancements in moving to MX and Flash Player 6 is the ability to import, edit, and play video. Now users are able to combine it with a level of interactivity such as allowing users to download a map of an exhibit hall onto their handheld devices.

DivX Web Player

DivX® <www.divx.com/divx/play/dwp/> has long been a popular movie download format, but streaming DivX video online was never realistic. DivX Web Player solves this problem by providing a browser plug-in to stream video online. Unlike past DivX efforts that were popular with Windows and took some time to filter into the Macintosh community, DivX now provides support for browser playback on both platforms simultaneously.

Pocket Player

Pocket Player <www.conduits.com/products/player/> is an alternative audio player for Pocket PC, featuring playback of common audio formats such as MP3, Ogg Vorbis, WMA, and WAV files. A ten-band audio equalizer and preamplifier control is supported, as is a visualization plug-in system. An auto-generated "Local Content" playlist, updated whenever content is added or removed, and an integrated playlist manager, means the user does not have to hunt for playlist files.

VITO SoundExplorer 2005

VITO SoundExplorer 2005 <http://vitotechnology.com/en/products/soundexplorer.html> is an MP3 player and audio recorder for Pocket PC that offers a comprehensive set of tools for school, professional, and hobby use. This software can turn a handheld computer into a universal multimedia digital dictaphone and MP3 player that provides comfortable and reliable mechanisms for interviewing, lecture recording, and listening to music.

PocketMusic Player Pro 3.71

The PocketMusic Player Bundle application <www.pocketgear.com/software_detail.asp?id=17133> can turn a Pocket PC into a portable media player with the ability to listen to MP3, OGG, MP1, MP2, WMA, and WAV music files from the Pocket PC main memory, the storage card (MultiMediaCard, Flash memory, SD card), the local network, or the Internet.

PocketStreamer Pro

PocketStreamer Pro: TV, Radio, Music, more 4.5 <www.pocketgear.com/software_detail.asp?id=18304&source=PK081605> is able to provide the Pocket PC with LIVE streaming TV, radio, music, news, and weather.

PocketTV

Pocket*TV*® <www.pockettv.com/> is a full-featured MPEG Movie Player for Microsoft Windows Mobile powered devices. Pocket*TV* provides outstanding video and audio quality along with many useful features not found in other video players.

On the new VGA-resolution devices (e.g. Dell Axim x51v, HP iPaq hx4700 series, Toshiba e800 series) Pocket*TV* takes advantage of the video hardware accelerators chips so it can play VGA-resolution MPEG files (i.e. DVD quality) with optimum quality.

Pocket*TV* works well with Pocket Internet Explorer, so users can play an MPEG movie by just tapping a link in a Web page. It is also capable of streaming MPEG video files using standard Internet protocols such as http, if the handheld device has a wireless network connection that supports the necessary bandwidth.

Pocket DVD Wizard 4.7

The Pocket DVD Wizard <www.pocketgear.com/software_detail.asp?id=13688> lets users convert DVDs and video files so they can download and play back on their Windows Mobile Pocket PC, Portable Media Center, or Windows CE device. All the conversion and file handling is done using the speed and power of the main Windows PC that provides excellent still frames and fast action video quality. This software uses advanced compression techniques to fit a whole DVD on a memory card the size of a postage stamp, resulting in a movie file that can be played on the standard Windows Media Player already installed on the Pocket PC.

TiVoToGo

TiVoToGo <www.tivo.com/1.2.15.asp> transfers TiVo-recorded shows to Portable Media Centers as well as smartphones and Pocket PCs with Windows Media Player 10 Mobile.

Subject Specific Software

Many subject specific software packages are available for classroom use. The leading group in promoting the use of handheld computers in the classroom is The Center for Highly Interactive Classrooms, Curricula & Computing in Education (The Hi-Ce) <www.hi-ce.org/>, a group of educators, computer scientists, psychologists, scientists, and learning specialists located in the School of Education at the University of Michigan. Their educational software can be accessed at <www.goknow.com/>.

Another leader in the field is K12 Handhelds, Inc. <www.k12handhelds.com/> that focuses exclusively on using technology in K-12 education. This company works with schools to provide professional development, hardware and software selection, on-going support, and follow-up for the successful handheld implementation into the curriculum.

Several online sites provide thousands of software packages that are suitable for educational purposes. EuroCool <www.eurocool.com/>, PalmGear <www.palmgear.com/>, and Palm Boulevard <www.palmblvd.com> are strictly Palm software sites. Pocket Gear <www.pocketgear.com/> and Pocket PC City <www.pocketpccity.com> only provide handheld software for the Pocket PC operating system, while Handango <www.handango.com/> provides software for both the Palm and Pocket PC.

Summary and Challenge

Software for handheld devices is relatively inexpensive and it may be worth the time to preview a single copy before ordering mass quantities. Readers for eBooks are often dependent on the eBook desired, so multiple readers may need to reside on a single handheld device. Likewise, several media players may also need to reside on a single handheld device, in order to be prepared to play the desired media without first stopping to download the specific media player needed. Specific Internet searches will need to be made to find subject specific software.

Before continuing, stop and consider the following questions: "What area in your curriculum, or what part of your collection, could be strengthened with the use of handheld computers?" "What media packages might provide the best eBook readers, eAudio, and eVideo players to meet the needs of your school?" After browsing the handheld software Internet sites ask, "What software will best meet the needs of your students, as well as the instructor and media specialists' needs?"

chapter four

Locating and Downloading Online eBooks

Electronic books, sometimes referred to as e-books, ebooks, or eBooks, are electronic versions of books, magazines, journals, reference manuals, textbooks, or any other documents traditionally occurring as a printed volume that can be viewed on a computer screen or handheld mobile device. Publishers, libraries, retailers, or private individuals can now produce, manage, and distribute eBooks to anyone on a computer running Microsoft Windows or Macintosh OS X.

History of eBooks

The first eBooks were referred to as hypertext fiction, a genre of electronic literature, which were non-linear and encouraged reader interaction. The reader typically chose links to move from one node of text to the next, which arranged a story in different sequences providing a deeper pool of potential stories.

The first hypertext fiction, "Afternoon," a story, was published in 1987 by American author Michael Joyce prior to the development of the World Wide Web, using software such as Storyspace and Hypercard®. In 1990, it was published on diskette and distributed in the same way as a book by Eastgate Systems.

The use of eBooks achieved serious recognition when the National Institute of Standards and Technology (NIST) and the National Information Standards Organization (NISO) held the world's first eBook conference in 1998. One of the big issues under discussion was eBook formats, which is still in contention today.

Since the stumbling beginning of digital publishing, The International Digital Publishing Forum (IDPF), formerly the Open eBook Forum (OeBF), was formed to include those with common goals to advance the competitiveness and exposure of digital publishing. This professional organization, which includes academic, trade, and professional publishers; hardware and software companies; digital content retailers; libraries; educational institutions; accessibility advocates; and other related organizations is responsible for setting standards and promoting the use of electronic publishing.

One of the first eBooks on the World Wide Web, *Riding the Bullet,* was published as a novella by Stephen King in 2000. It was available as a free download for the first week or so as a publicity stunt. As a result, the load on the Web servers hosting the story was so high that it rendered them totally inaccessible. In 2002, the story was published as part of the collection *Everything's Eventual: 14 Dark Tales.*

Hazards of eBooks

"Do not judge a book by its cover" is an old adage that is just as relevant today when purchasing eBooks as it was for a print book. An attractive cover may look professional but it does not guarantee good writing. Artistic covers can merely be a mask, since in today's digital age an appealing picture can be generated quickly and effortlessly using an image-editing tool. Many self-published eBooks do not go through a copy editor or peer review, and they may be poorly written and contain factual errors.

Before downloading an eBook, copyright and distribution rights need to be considered. Educators need to check before purchasing an eBook to see if that book is available free elsewhere. Many classic books, pre-1920s, have passed into the public domain and now are available for download from one of the free online libraries. However, it is surprising how many people believe that because they find a copy of *Harry Potter, Lord of the Rings*, or other popular eBooks on the Internet, they assume the publisher released it for electronic distribution. In many cases, publishers do not often release the digital rights to recent, popular books.

While eBooks are relatively secure, they can still be mass distributed much faster than a paper book, depriving the creators of their legitimate earnings. Therefore, before downloading a book to do research, if it is a popular book, search for it on a major eBook site. If it cannot be found or if there is not any news about the book's release for electronic format, chances are it may not be legitimate.

Educational Advantages for eBooks

One of the greatest advantages of students using eBooks is the size of the eBook. The weight of student backpacks has long been a concern of pediatricians, parents, and teachers. According to Dr. Marvin Arnsdorff in "Mounting Research on Backpack Use" originally published in I.C.P.A. Newsletter May-June 2002 <http://www.icpa4kids.org/research/articles/childhood/backpack_research_newsletter.htm>, the proper maximum weight for loaded backpacks should not exceed 15 percent of the child's body weight. This means that an 80-pound child should not carry more than 12 pounds in a pack.

In place of a 20-pound backpack, a student can now carry a simple handheld computer weighing less than 14 ounces that contains scores of books, articles, and reference works plus a means to record notes while reading those articles. Teachers can add notes, organizers, comments, and questions to sections of those books to help direct the student's studies.

The main reasons for using electronic materials include currency of the material, price, availability, and access to materials no longer in print. With the high cost of print textbooks and the fast changing information to be included in those books, eTextbooks have become a viable option for price and currency. Companies that offer electronic versions of textbooks include Holt, Rinehart, and Winston <http://hrw.com/it/index.htm>, Pearson Prentice Hall <www.phschool.com>, and McGraw-Hill <http://mhln.com>.

An article entitled "Study: States are slowly embracing eTexts" in *eSchool News Online* By Robert Brumfield, Assistant Editor, eSchool News dated September 28, 2005, <www.eschoolnews.com/news/showStory.cfm?ArticleID=5883> states:

> "Nearly all states with textbook adoption policies now include software, digital content, and other technology-based media in their definitions of 'instructional materials,' according to a recent survey by the Software and Information Industry Association (SIIA). But only a third have updated their submission or review processes to account for unique technology issues not otherwise faced with printed textbooks. Nearly all states with textbook adoption policies now include software, digital content, and other technology-based media in their definitions of 'instructional materials,' according to a recent survey by the Software and Information Industry Association (SIIA). But only a third has updated their submission or review processes to account for unique technology issues not otherwise faced with printed textbooks."

Transferring eBooks from Desktop to Handheld Devices

HotSync is the process of synchronizing information between a Palm handheld device and a desktop or laptop computer. A press of the HotSync button on the handheld's cradle (a dock station) usually activates the HotSync process. This application communicates with various conduits on the desktop PC to install software, backup databases, or merge changes made on the PC or the handheld to both devices.

In addition to the conduits provided by the licensee, developers can create their own conduits for integration with other handheld applications and desktop products. A Backup conduit included with the HotSync software backs up (and restores, if necessary) most of the data on a Palm handheld device. There are currently tens of thousands of third-party applications available for the handheld operating platforms, which have various licensing types, including open-source, freeware, shareware, and traditional commercial applications.

eBook Formats

Plain Text

Plain Text (.txt) eBooks can be read by any text reader or word processor. One freeware program, Tom's eTextReader <http://pws.prserv.net/Fellner/Software/eTR.htm>, allows users to read plain text files such as those provided by Project Gutenberg—in a book-like manner. Users can select window size, font style, and font size while the page breaks are inserted automatically. To add to the book's educational value, readers can also set bookmarks, find words or phrases, edit the text with an internal text editor, or use the 'Find text in files' tool.

Adobe Acrobat Format

Adobe® Acrobat® format (.pdf) for Pocket PC Windows Mobile OS or for the Palm OS <www.adobe.com/products/acrobat/readstep2.html> extends the value and capabilities of Adobe PDF files by adapting the files for high-quality viewing on smaller screens, while preserving their rich content. Adobe Reader makes it easier to import, display, and share digital editions on handheld devices and provides possibilities for wireless printing. Digital pictures can also be displayed or shared by using the Adobe Photoshop Album slide shows. Read Out Loud is a Text-to-Speech (TTS) tool that is built into Adobe Reader 6.0 <www.adobe.com/enterprise/accessibility/reader6/sec2.html>. This feature reads text contained within a document window.

Adobe Content server, a digital rights management (DRM) software, provides tools, storage capacity, and tracking systems to enable libraries to lend eBooks and other electronic documents as secure, reliable Adobe PDF files. This Adobe eBook technology allows library patrons to borrow, download, and read offline a wide variety of titles on any computer running Microsoft Windows or Macintosh OS or download the title to a Palm or Pocket PC handheld device. Once a book is checked out, it can be accessed for a set amount of time, after which it is automatically deactivated, making it available for other patrons.

From the reader's point of view, the Adobe PDF eBook provides a digital-reading experience complete with a dictionary, Web links, and graphics. Adobe Reader® for the handheld computer conveniently reformats Adobe PDF text to fit and to be easily read on small screens, while preserving graphics and images. Adobe Reader also has the capabilities to display Adobe Photoshop Album slide shows.

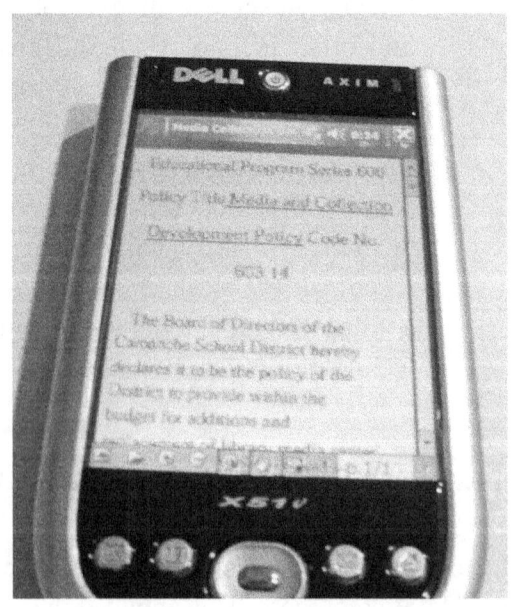

Figure 4:1 Adobe Reader on a Handheld Computer

MobiPocket Reader

MobiPocket Reader (.prc) <http://mobipocket.com/> is a French company incorporated in March 2000. MobiPocket is a universal eBook reader for handheld computers that is freely downloadable. It contains publishing and reading tools for both the Windows Mobile and Palm operating systems. Their latest reader contains a major upgrade of the Web Companion desktop software. Users can now subscribe to eNews from many major Web sites and synchronize them directly to the memory card of a handheld or smartphone. Microsoft Word, Excel, PowerPoint, and other Office documents can be converted to the Mobi .prc file format with the MobiPocket Office Companion <www.mobipocket.com/en/DownloadSoft/DownLoadOfficeCompanionStep1.asp>.

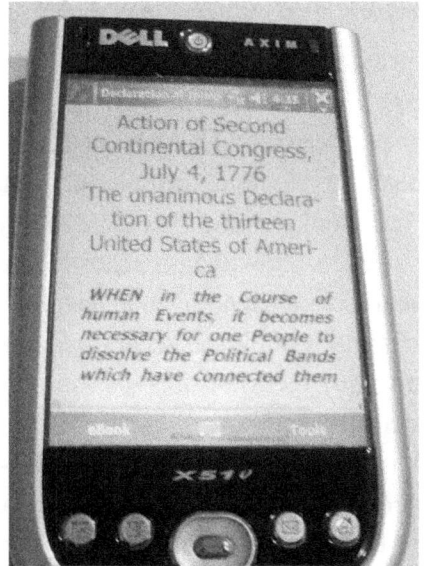

Figure 4:2 MobiPocket Reader

Microsoft Reader

Microsoft Reader (.lit) <www.microsoft.com/reader/> is a free software application that includes patented Microsoft ClearType® display technology. This format is extremely secure and the most professional looking among the eBook readers. ClearType technology improves resolution on LCD screens by up to 300 percent to deliver a print-like display. Microsoft Reader offers a clean, uncluttered layout with ample margins, proper spacing, and leading, plus powerful tools for marking, highlighting, and annotating the eBooks while reading.

Microsoft Reader is able to take advantage of existing speech technologies by simply installing Microsoft Reader Text-to-Speech (TTS) Package, <www.microsoft.com/reader/developers/downloads/tts.asp>, which enables voiced reading and navigation features for Microsoft Reader on Windows-based PCs and laptops in English, French, or German.

The downside of the Microsoft Reader is that it is not available for the Palm operating system.

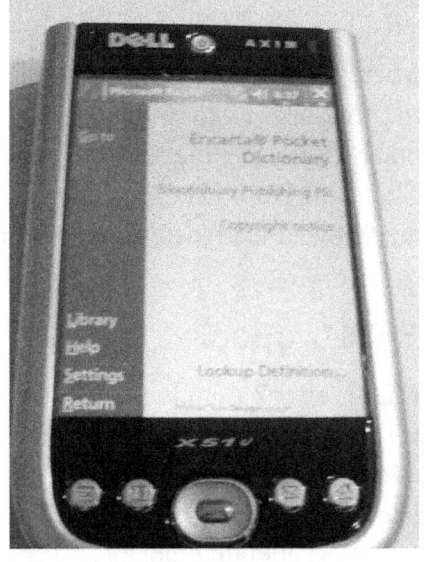

Figure 4:3 Microsoft Reader

eReader and eReader Pro for Palm OS

eReader and eReader Pro for Palm OS (.pdb) <www.ereader.com/product/browse/software> allows users of virtually any platform to read eBooks. Using an eReader allows users to add bookmarks, easily navigate an eBook, and search for individual words within the eBook. There is no charge to download and use free eReader software. However, a feature-enhanced eReader Pro is available that provides additional reference books. For example, while reading a book the reader can select any word and easily look it up using the Merriam-Webster's Pocket Dictionary that is included with the software.

Using eReader Pro allows users to customize and create color themes and adjust several display features such as justification of text and line spacing to personalize the eReader. This program also contains Autoscroll viewing features to enhance readability. The included Agfa Monotype Fonts contain three font faces for eBook reading as well as supporting four sizes for each font.

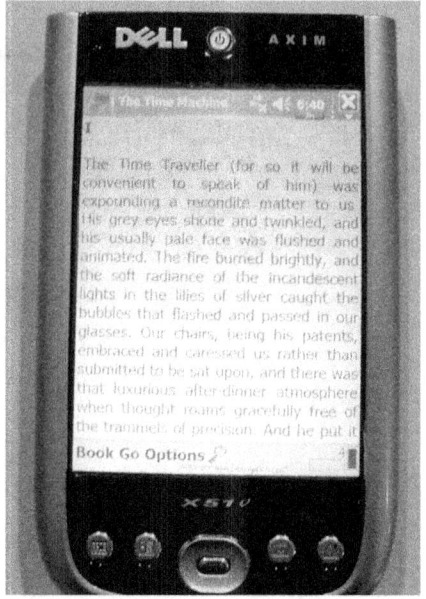

Figure 4:4 eReader Pro

Microsoft Word eBook

Microsoft Word eBook (.doc) provides e-reading in a familiar setting—Microsoft Word. If users have Microsoft Word installed on their computer, they will not need additional software. They can change the font size as they would any other Word document, and can print an eBook as they would any other Word document. Most publishers do not offer their eBooks in Microsoft Word format because Microsoft Word eBooks are not "secure" programs, like Microsoft Reader or Adobe Reader. Customers are allowed to transfer their Microsoft Word eBooks to additional computers with no restrictions.

HTML eBook

HTML eBook (.html/xml) can be used because an eBook is similar to a long Web page. Users can read HTML eBooks on their Web browser, such as Internet Explorer, Mozilla's Firefox®, or Netscape®, but the eBook is stored on the computer instead of the Internet. No additional software is needed except a browser, like Internet Explorer or Netscape Navigator®. Many handheld devices require an extra reader such as MobiPocket <http://mobipocket.com/> to read HTML eBooks. Once a user has purchased and downloaded an eBook, it can be opened and read in a browser any time—even when it is not connected to the Internet. HTML eBooks can also contain hypertext navigation, just like what users are familiar with on Web pages.

Sources for eBooks

- Michael Hart, the inventor of the eBook, founded Project Gutenberg <www.gutenberg.org/> in 1971. It is currently the largest single collection of free electronic books and makes nearly any public domain work available online in digital format. Project Gutenberg claims that nearly 2 million eBooks are downloaded each month.

- eBooks.com <http://ebooks.com/> sells whole books, chapters, and pages of 45,000 popular, professional, and academic eBooks online from the world's leading publishers. Usually eBooks can be purchased and downloaded immediately by customers anywhere in the world. In partnership with leading university and research libraries and major academic publishers, eBooks Corporation recently launched EBook Library (EBL), an eBook lending platform that provides a wide selection of eBooks to academic and research libraries.

- Digital Book Index <www.digitalbookindex.org/> provides links to more than 105,000 title records from more than 1,800 commercial and non-commercial publishers, universities, and various private sites. About 66,000 of these books, texts, and documents are free, while others are available at low cost.

- eFollett.com eBookstore <http://ebooks.efollett.com/> is a network of bookstores serving campuses across the United States and Canada. They have a wide variety of subjects, including medical, reference, and even popular fiction. Users can download eBooks and read them on their computer or transfer to their mobile device.

- Microsoft Reader eBooks <www.mslit.com/> provides inexpensive, popular eBooks. Users can read the same eBook on both their desktop and their Pocket PC by activating the Pocket PC.

- International Children's Digital Library <www.icdlbooks.org> is a public library for the world's children sponsored by the University of Maryland. The library contains nearly 1,000 free children's books written in more than 33 different languages.

- MemoWare <www.memoware.com> contains a collection of 18,000 documents, databases, literature, maps, technical references, and lists.

- Fictionwise™ <www.fictionwise.com/> is a leading independent eBook publisher and distributor that publishes award-winning and high-quality eBooks by top authors in multiple formats.

- Wikibooks <http://en.wikibooks.org/wiki/Main_Page> is an attempt to create a comprehensive, kindergarten-to-college curriculum of textbooks that are free and freely distributable, based on an open-source development model. Wikibooks is created in the same mold as the Wikipedia project—the open-source encyclopedia that lets anyone create or edit an article.

- Bartleby Bookstore <www.bartleby.com/> is an Internet publisher of literature, reference, and verse for both students and researchers.

- The Alex Catalogue of Electronic Texts <www.infomotions.com/alex/> is a collection of public domain documents from American and English literature as well as Western philosophy.

- Amazon e-Documents <www.amazon.com/exec/obidos/tg/browse/-/551440/102-6117958-3759301> carries more than one million digital documents that are delivered to the users Digital Locker immediately upon purchase. Amazon also began its Search Inside the Book (SITB) program in October 2003. This program provides limited full text and full image access to hundreds of thousands of books available through Amazon.com. The SITB consists of two different programs, Amazon Pages and Amazon Upgrade, each intended to complement the other. Amazon Pages lets users buy one page at a time. The publishers who are the copyright holders will get to make the decisions about how much each page costs. Users can buy a page, or a chapter, or possibly assemble their own textbook. With Amazon Upgrade, users must buy the physical book. When they buy the physical book, they also can buy online access and obtain instant access.

- Google Print Library Project <http://print.google.com/> wants to digitize and make searchable online texts from the university library collections at Harvard, Stanford, Michigan, Oxford, and the New York Public Library. However, this project has been fraught with criticism. The Authors Guild and the Association of American Publishers sued Google over its Google Print Library Project arguing that scanning entire copyright works violates copyright law. Google claimed it was abiding by the "fair use" provision of the copyright law.

- Planet PDF is a free PDF eBooks archive <www.planetpdf.com/free_pdf_ebooks.asp>.

- Bibliomania <www.bibliomania.com/> includes more than 2,000 texts of classic literature, book notes, references, and resources in html format.

- CIA Publication Library <https://www.cia.gov/cia/publications/index.html> includes the World Fact Book with information on every country in the world.

- Manybooks.net <http://manybooks.net/> contains more than 10,000 eBooks Palm, Pocket PC, Zaurus, Rocketbook, or PDA in multiple formats, including versions for the iPod.

- Baen Free Library <www.baen.com/library/> features science fiction books.

- Inebooks.com <http://inebooks.com/> produces and sells interactive children's eBooks for PDAs. Each interactive book doles out short passages of story and then allows the reader to make decisions about how the story progresses. The format is quite similar to the Choose Your Own Adventure line of books made popular in the early '80s.

- eBooks-in-Print.com <http://ebooks-in-print.com/> is a free, up-to-date, searchable, online catalog that lists registered titles available from participating electronic publishers. It provides links to publishers and other download sites.

College and University Electronic Book Collections

Universities around the world have long been interested in digitizing research materials. With a little research, eBooks that are suitable for K-12 curriculum are available in these collections. The most well-know include:

- University of Michigan's School of Information's Internet Public Library <www.ipl.org/> claims to have links to more than 40,000 eBooks that can be read online or downloaded free of charge.

- University of Pennsylvania's Union Catalog of eBooks <http://onlinebooks.library.upenn.edu/> is an independent online book repository that facilitates access to books that are readable over the Internet. They encourage the development of online books for the benefit and edification of all readers.

- University of California Press Books eScholarship Editions <http://content.cdlib.org/ucpress/> includes almost 2,000 books from academic presses on a range of topics, including art, science, history, music, religion, and fiction. Access to the electronic books is open to all University of California faculty, staff, and students, while select books are available to the public.

- University of Texas at Austin, Electronic Books <www.lib.utexas.edu/books/etext.html> provides a large selection of digital books for its faculty, staff, and students through numerous original sources. They also provide links to numerous other digital collections.

- University of Virginia Electronic Text Center <http://etext.lib.virginia.edu/ebooks/> provides access to more than 2,100 eBooks. Their collection includes classic British and American fiction, major authors, children's literature, American history, Shakespeare, African-American documents, the Bible, and much more.

eBooks for People with Disabilities

Assistive technology is one of the most educationally sound uses of technology available. Making eBooks available to students with disabilities can be a vital tool for both the classroom teacher and the library media specialist. Sources for eBooks for the disabled include:

1. Bookshare.org <www.bookshare.org/> is a Web-based system supplying accessible books in digital formats designed for people with disabilities. These digital formats are the NISO/DAISY XML-based format for the next generation of talking books, and the BRF format for Braille devices and printers. Access to copyrighted books from Bookshare.org is limited to people in the United States with bona fide print disabilities and the schools and nonprofit organizations serving them. Bookshare.org takes advantage of a special exemption in the U.S. copyright law that permits the reproduction of publications into specialized formats for the disabled.

2. Accessible Book Collection <www.accessiblebookcollection.org/> offers an eBook subscription service for readers with special needs.

3. The Texas Text Exchange <http://tte.tamu.edu/> is the first Web-based digital library of electronic books for exclusive use by students with disabilities.

4. Visibooks <http://visibooks.com/> are illustration-based computer class textbooks that contain one-tenth as many words as conventional computer texts. Visibooks were developed as part of a U.S. Department of Education research study on textbooks for students with learning disabilities and/or low English proficiency students. The study showed that the books worked well for everyone, not just special needs students.

Specialized eBook Search Engines

Other sites to consider when searching for eBooks that will best meet the students' and instructors' needs include specialized eBook search engines. Such sites include:

1. Digital Book Index <www.digitalbookindex.com/search001a.htm> indexes most major eBook sites, along with thousands of smaller specialized sites.

2. EBookLocator <www.ebooklocator.com/> searches a database of books available at leading retail sites and public library sites.

3. Google Scholar <http://scholar.google.com/> provides a tool to search scholarly papers.

4. Google Book Search <http://books.google.com/> is a Google digitized book service.

5. John Labovitz's E-Zine-List provides a list of electronic 'zines <www.e-zine-list.com/> around the world, accessible via the Web, FTP, email, and other services.

Summary and Challenge

Millions of eBooks are available in all genres, subject matters, reading levels and quality levels, and even for people with disabilities. By taking advantage of the variety of eBook formats for portable devices, students, teachers, and librarians can increase the accessibility to information regardless of where they are. The ease of this accessibility enhances student learning as it improves their reading skills while the portability encourages better use of time management. This in turn will lead to an increase in reading scores.

Before continuing, ask yourself these questions: "Where and how will you search for eBooks for yourself and your students?" and "What would be the best ways for your students to search for their own eBooks?"

chapter five

Accessing Web Sites on Handheld Computers

Students and educators often need access to the Internet for research or study from a variety of locations without the availability of a desktop or laptop computer. Handheld computers can be an obvious solution. Both the Palm OS and the Windows Mobile operating systems come on handhelds, with either wireless or Bluetooth technology for Internet accessibility, that allow users to surf the Internet in real time. Most handheld devices allow users to download Web sites from a desktop and read at the user's convenience.

Mobile Browsing

Not all Web sites are optimized for mobile browsing. The main difference between a mobile Web site and a desktop/laptop Web site is that the mobile site information is condensed into a single column with fewer graphics. Because the addresses for mobile sites differ from the addresses used for desktop computers, a number of search engines have compiled a list of mobile-enabled sites that can be accessed from a handheld computer or cell phone.

To locate mobile services and sites easily, users can take advantage of almost any search engine and search for "mobile sites." The most popular mobile search tools are Yahoo!® Mobile <http://mobile.yahoo.com/> or Google™ Mobile <http://mobile.google.com/mobile_search.html>.

Another method of obtaining a list of mobile sites is to use a portal site with links to sites that are grouped by category for mobile devices with these links. Such portals include:

- eboogie.com <www.eboogie.com/> contains a list of mobile device friendly home pages.

- Elghazi.com <www.elghazi.com/mobile/> contains a mobile portal.

- Fleximo <www.fleximo.com/> explores the mobile Web and finds other portals with mobile sites.

- HotBot <www.hotbot.com/> searches the Web using Ask Jeeves or Google. Wireless iMenu <www.imenu.com/wireless> provides a menu of links to pages such as Google, Yahoo!, and numerous publications.

- InfoSpace <www.infospace.com/info.avant/infoapp> links to helpful resources, including white and yellow pages, reverse phone lookup, maps, city guides, and more.

- MobileLeap <http://mobileleap.net> detects the device type, and instantly reformats Web sites by compressing data, filtering out unsupported content, and fitting the pages to the mobile device screen.

- O2 Mobile Portal <www.seeo2.com/mobile/template/MobileMain.vm/> accesses news, travel, and entertainment sites.

- PDA Hotspots <www.wacklepedia.com/pdahotspots/pda_hotspots.htm> claims to have 1,000 links to Web sites compatible with both Palm and Pocket PC. PDAportal.com <http://pdaportal.com/> contains a portal for the small screen.

- PDA Tree <http://pdatree.com> contains links to mobile Web sites. PdaMobileWeb <http://www.pdamobileweb.com/> searches the Web and browses a variety of mobile sites organized by category.

- PDAntic Portal <www.pdantic.com/pdaportal.htm> is designed for wireless devices in categories ranging from entertainment to finance to weather.

- PliNkIT! <www.plinkit.com/> includes links to categories, including life, entertainment, travel, sports, and business.

- Skweezer <www.skweezer.net/> contains a directory of sites, but also lets users "skweeze" any Web site to view it on their mobile device.

- Wapedia Mobile Encyclopedia <http://pda.en.wapedia.org> makes the content of Wikipedia available for viewing on mobile devices.

- Wxnation <www.wxnation.com/wireless> is a site from Fort Worth, Texas, that provides local weather, traffic, and live cam resources along with links to other mobile weather sites.

Using Mobile Favorites

If a direct Internet connection is not available, Web sites can be downloaded onto a handheld computer when it is synchronized with the desktop computer. Mobile Favorites are built into the operating system of most handheld devices so that users can download a Web site and read it off-line on their handheld device.

Users rarely think about it when they surf the Web on their desktop computers, but on a handheld device entering a URL into a mobile browser becomes tedious and more error prone when the user is entering data with the on-screen keyboard or thumb-typing.

A great deal of time and frustrations can be saved by selecting *Favorites*, a list of desired Web sites for the handheld computer that is located on the toolbar of Internet Explorer on the desktop computer, and then synchronize those Web sites to a handheld computer. Each manufacturer has different set-up rules, so each handheld model will differ slightly, but the basics are nearly the same. Individual help menus and manuals may need to be consulted for details.

The first step is to connect the handheld to the desktop computer the same as if ActiveSync will be running. Since not all Web sites are optimized for mobile browsing, check to see if the desired site is mobile-ready. If it is mobile ready, the site can be added to the Mobile Favorites via Internet Explorer by clicking the PDA icon in the Standard toolbar. Some sites that are not mobile-ready can be viewed reasonably well on a handheld device. It may be worth downloading the site and deleting it from the handheld if it is too difficult to read.

To download Web sites to mobile devices using Internet Mobile Favorites, the handheld computer needs to be connected to the Desktop or laptop PC:

Figure 5:1 Screenshot of Explorer Toolbar

- Using Microsoft Internet Explorer, locate the Web site on the desktop or laptop computer.

- Click the Mobile Favorite Icon on the Explorer Toolbar and a "Create Mobile Favorite" pop-up screen appears.

- Name the Web site and press the "Create In" button.

- Save the link in the desired folder, such as "Mobile Favorite," and press "Okay."

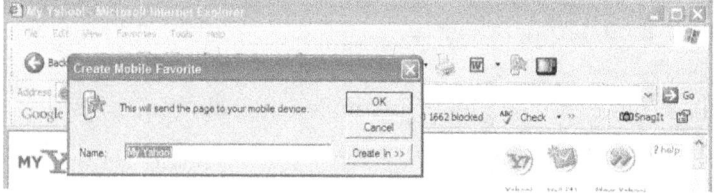

Figure 5:2 Screenshot of Pop-up Screen

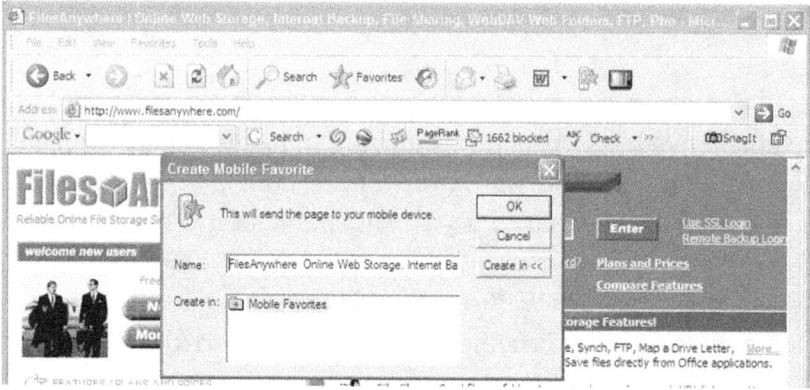

Figure 5:3 Screenshot of Next Pop-up Screen

To save two or three layers of a Web site to mobile devices using Internet Mobile Favorites, the handheld computer needs to be connected to the desktop or laptop PC. Care must be taken in the number of layers downloaded as it could overwhelm the storage space of the handheld.

- Under Favorites on the toolbar, select "Organize."

- Select the desired URL with the Mobile Favorites Folder.

- Check the box "Make Available Offline."

- Click on "Properties" box.

- Select the "Download" tab on the top.

- Change the 0 to 1, 2, or 3 in the Download links deep box.

- Select "Apply" and "Okay."

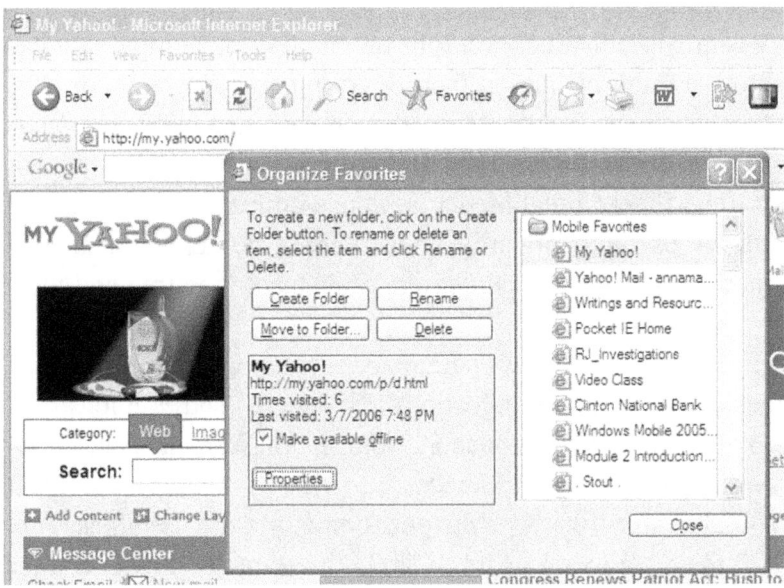

Figure 5:4 Screenshot of Organize Box

When the selection of Mobile Favorites is completed on the desktop computer, they will be loaded onto the handheld computer the next time the handheld is synced with that desktop. To access the Mobile Favorites on the handheld device: Tap "Start," Tap "Internet Explorer," Tap the Favorites icon (Yellow Star) located near the bottom the screen, Tap the Mobile Favorite link, and the Web site will load.

To synchronize the settings manually from the Internet to the desktop system, click on the Internet Explorer Favorites button to show the Favorites pane on the left hand side of the screen. Select Mobile Favorites, then right click on the favorite for link, and select Synchronize.

Web Clipping Software

Web clippers are commercial products that are Internet Explorer add-ons. They are available for both Windows and Macintosh desktop computers and have handheld readers for both Palm and Windows Mobile operating systems. After installing the Web clippers on Windows desktop computers, they create toolbar buttons for Internet Explorer.

Namo HandStory Suite 3.1 <www.namo.com/products/handstory/> is a multifunction information software browser and Web clipper for the Palm or Windows Mobile operating systems, and is able to displays texts, eBooks, Images, Web Clips, and even Video Clips (for Pocket PC only) all from within a single application. Namo HandStory Converter for the PC is a versatile converter of texts, images, and Web content. HandStory Media Suite Browser can play video files that have been converted from standard video file formats using HandStory Converter on a Windows PC or downloaded from the Internet.

iSiloX <www.iSiloX.com> is used for clipping entire articles from Internet Explorer and converting the content to the iSilo 3.x/4.x document format, enabling users to carry the article on a Palm or Pocket PC handheld device. Users like iSiloX because it is highly configurable, can shrink images, and reformat tables to single columns. The ability to shrink images and reformat tables means users are not limited to pages specially formatted for the small screen of a handheld device. iSiloX lets users create formatted, easy-to-read, compressed documents that can then be read using the iSilo reader.

WebCopier for Pocket PC™ <www.maximumsoft.com/> runs on the desktop computer. It can download Web sites, so that users can browse them offline on their desktop computer, and copy the sites to their Pocket PC. Pocket WebCopier users can browse Web sites on their handheld computers at their convenience.

Sync & Go is part of Microsoft's $19.95 Plus! Digital Media Edition, a multimedia enhancement pack for Windows XP. Sync & Go allows users to download multimedia content and have it automatically synced to a Pocket PC. Once the software is purchased, downloading is free and there are no charges for premium content.

MobiPocket Reader <www.mobipocket.com/en/eNews/default.asp> contains an added feature that allows users to read eNews Web sites, Real Simple Syndication (RSS) feeds, and Weblogs on a handheld computer or smartphone. It provides 500 eNews channels from major Web sites and syncs the selected ones directly to the memory card of a handheld device.

Sunnysoft World Off-Line (WOL) <http://www.pocketgear.com/software_detail.asp?id=5797/> is an application for Pocket PC 2002 and Windows Mobile 2003 that directly downloads the content of any Internet page to the handheld device. World Off-Line supports the download of images, sub-pages, and files as well as the off-site content download. Users can customize what and when to download to the handheld device. An important advantage in comparison to AvantGo is that World Off-Line provides a direct connection between the handheld device and the Web content, without a third party participating in the download process.

AvantGo

AvantGo <www.avantgo.com/> is a free service that provides access to more than 1,000 Web sites optimized for mobile devices. AvantGo can be used with a desktop computer or wireless Internet connection. AvantGo offers thousands of channels from the world's leading brands in news, weather, sports, and more. It can also be used to synchronize any Web site (family site, workout schedule) to a handheld device.

AvantGo can be a convenient mobile travel assistant that provides access to a users personal travel itineraries from leading airlines and travel services—plus city guides, weather, maps and directions, and more.

Streaming Digital Media via the Web from a Desktop to a Handheld Device

Handheld users can access the media on their desktop computer anywhere there is a wireless network using a simple free program called Orb <www.orb.com>. In addition to streaming music, Orb can send live television, digital photos, home video, and recorded TV programs to a Pocket PC. Orb works with Windows XP Media Center edition PC or a standard Windows XP PC. Orb can stream media content via an agent that is installed on a personal PC and then through any Web browser, including the one on a Pocket PC. Users can even access their Web Cam through Orb and monitor activities at another location.

Summary and Challenge

As more and more schools are becoming wireless, and more hot spots are becoming available in the community, students will have an ever-increasing opportunity to take advantage of immediate Web access. The time and resources saved in having such ease of access can be applied to a multitude of educational endeavors and students can work much more productively.

Even if a wireless Internet connection is not available, with a minimum of planning students and teachers can select the necessary Web sites, download, and synchronize them to the handheld device to be referenced later. For example, the time students spend on a bus or waiting for a bus might be used to research topics for the school curriculum on their handheld devices. Wasted minutes can be transformed to learning minutes.

Before continuing, ask yourself these questions: "What Web sites would best enrich your curriculum?" "Can they be easily viewed on the smaller screen of a handheld device?" "How can students access those sites?" and "Will the Web sites need to be downloaded or do the mobile devices have Internet access and the site can be read online?"

Curricular Uses of Student Produced eBooks

Not only can students write eBooks for language arts classes, but also eBooks can be used in a similar fashion for foreign language classes. Students can collaborate in the development of an eBook, make changes on the eBook, and then beam it to their classmate for them to make changes.

Research papers in science and social studies can be prepared in the eBook format. Research can be done directly from the handheld or it can be done on a desktop or laptop and synchronized onto the handheld device. Students can then collaborate via the handheld devices to obtain one finished, polished product.

Not only can students prepare their own eBooks, but teachers can also use the same tools to prepare assignments specifically for the needs of individual students. Educators can add notes, organizers, questions, and comments to text before converting it to an eBook format for students. Teachers and media specialists can also use student-created eBooks to create digital libraries of student work that can be shared with the class or with the world by publishing the students' work through a class or school Web site.

eBook Writers

Many tools are available for creating text-format eBooks, including common application software such as Microsoft Word as well as several free programs dedicated to creating or converting existing electronic text materials into eBooks. Each has its own strengths and weaknesses.

The first step in eBook creation is preparing the content. For original projects, the creator needs to type the text into a text editor or import text using a scanner running optical character recognition (OCR) software such as TextBridge® Pro 11 <www.nuance.com/textbridge/> for Windows Operating Systems or Readiris® Pro 11 <www.irisusa.com/products/readiris/MAC/> for Macintosh.

Plain Text eBook

The simplest form of an eBook is the plain text eBook in either the text (TXT) or rich text format (RFT). Any word processor can create text eBooks by saving the file in either text or rich text format. Saving the eBook as a plain text file means that the eBook will contain only unformatted text, not pictures or special font formatting. To include graphics and other features, the file needs to be saved in rich text format. Text eBooks can be read with word processors, Web browsers, and special eBook readers, such as Tom's eTextReader for the PC.

HTML eBooks

HTML eBooks can be easily created using a Web editor such as Dreamweaver®, Microsoft FrontPage®, Netscape Communicator, or any number of other Web Editors. HTML eBooks can also be created by using a word processor program by using the Save as HTML or Save as Web Page options.

Unless a user has a program such as Adobe Acrobat, <www.adobe.com/products/acrobatstd/main.html> users will need to create the eBook material in one format and convert it to PDF format. Some word processors, such as OpenOffice <www.openoffice.org/>, allow users to save documents as PDF files. If the word processor cannot create PDF files directly, an online converter, such as the one provided by BCL Technologies <www.gobcl.com/convert_pdf.asp> can be used. This Web site runs a program, converts the file, zip compresses the created files, and emails the file to the given address.

Microsoft Reader eBooks

Creating eBooks for Microsoft Reader is a straightforward process using Word 2000 or above. The Microsoft Reader (RMR) plug-in <www.microsoft.com/reader/developers/downloads/rmr.asp> needs to be first downloaded and installed. A new listing of READ is added to Word's FILE menu, which will temporarily convert a Word document into a Web page and then into an MS Reader eBook. When the pop-up window appears, the desired specifics must by designated and when the "Convert to Microsoft Reader Formatting" box is checked, Word automatically removes formatting that conflicts with Microsoft Reader defaults. The MRM plug-in automatically generates a linked table of contents from the table of contents in the Word document. For assistance in generating a table of contents in a Word document, one may need to refer to the Microsoft Word help menu.

Any graphics or images included in the original Word document will be transferred into the body of the eBook. However, the Reader formatting mode needs to be used to avoid potential image handling errors. To include the original cover graphics in an eBook, click the "Customize Covers . . ." button at the bottom of the RMR window.

For foreign language and English as a second language teachers, Microsoft Reader can create eBooks for any language using European characters. Microsoft Reader will automatically hyphenate only English, French, German, Italian, and Spanish text correctly.

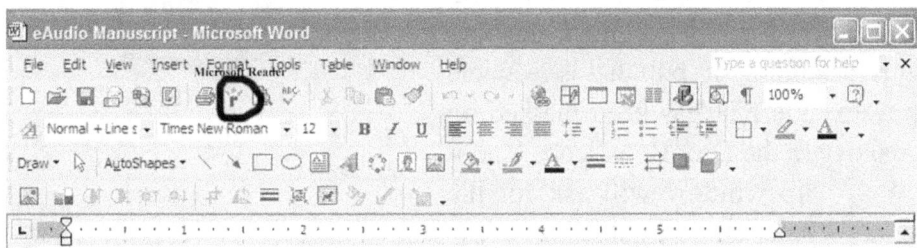

Figure 6:1 Screenshot of Microsoft Reader Plug-in

ReaderWorks

Another program, ReaderWorks® <www.overdrive.com/readerworks/>, also creates eBooks in the MS Reader (LIT) format. The standard version is a free download while the professional version is a full-featured eBook authoring and conversion package for commercial development. It includes all the easy-to-use tools of ReaderWorks Standard and has enhanced features enabling the customization of an eBook for commercial distribution and sale.

ReaderWorks Publisher contains a Table of Contents Wizard that automatically scans the content pages and generates a table of contents with built-in hot links. The Table of Contents Wizard also permits users to customize the appearance of the table of contents, create their own styles, and preview their table of contents before building the eBook.

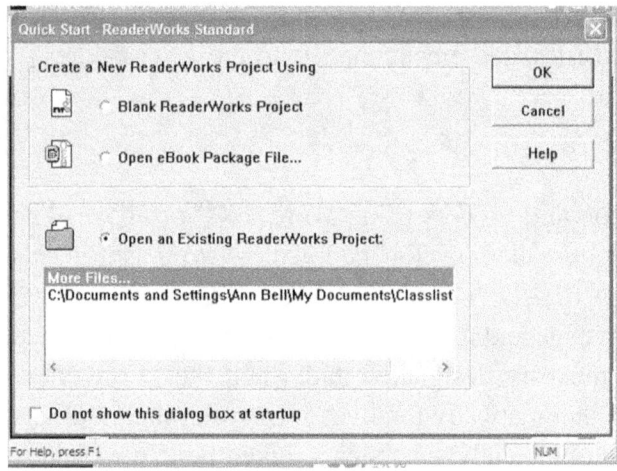

Figure 6:2 Screenshot of ReaderWorks Program

The 2.0 release of ReaderWorks supports building a Microsoft Reader eBook title from either HTML files, Word document (Word 2000 and higher), an Open eBook package file, or ASCII text. Most popular desktop publishing and word processing programs can export into one or more of these formats.

ReaderWorks can be run on a Macintosh computer only if the Macintosh is running Virtual PC; however, Microsoft Reader cannot be run on a Macintosh. Therefore, Microsoft Reader eBooks can be created on a Macintosh, but will not be able to be read on either a Macintosh or a Palm Operating System.

DropBook

The free program DropBook <http://ebooks.palm.com/dropbook> can convert a text file coded with Palm Markup Language (PML) tags into an eReader eBook for use on Pocket PC and Palm handheld devices as well as Windows and Macintosh desktop computers. To create a file with the ending of .pdb that is readable by the eReader, simply drop a text file that has been marked up using the Palm Markup Language onto the DropBook icon. A pop-up window will ask for the title, author, publisher, copyright, and ISBN number.

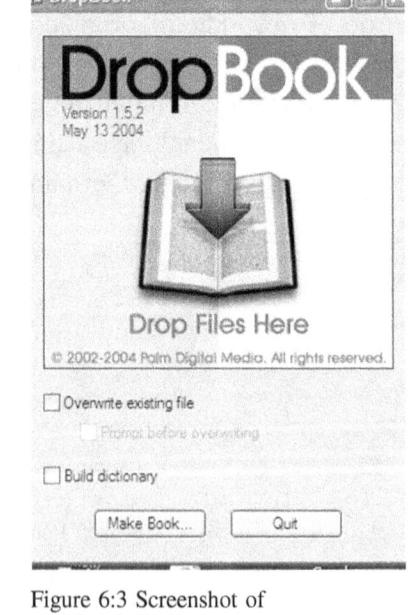

Figure 6:3 Screenshot of DropBook Program

eBook Studio

Another excellent program for creating eBooks is eBook Studio <www.ereader.com/products/ebookstudio>, which creates eBooks that can be read by the eReader and eReader Pro software on either the Palm or Pocket PC handheld devices. With eBook Studio, users will be able to design the navigation of the text, format it, create chapters/table of contents, add images, create links, and convert the finished document for reading on a handheld device.

Palm eBook Studio supports both Windows and Macintosh operating systems. eBook Studio provides a desktop What-You-

Figure 6:4 Screenshot of Palm eBook Studio Program

See-Is-What-You Get (WYSIWYG) editor that allows users to lay out their eBook exactly as it will be on the handheld. A free demo is available and the full version is currently $29.95. However, eBook provides a School Purchase Program to assist teachers and administrators in deploying eBook technology in classrooms, large school projects, and district-wide initiatives.

Mobipocket Creator

Mobipocket Creator 4.0 <www.mobipocket.com/en/DownloadSoft/> allows users to create an eBook by adding HTML pages, Word documents, and image files to the Publication List. Creators can then build and test their eBook on the PC with the built-in Emulator and send the eBook to a handheld device with only a few clicks of the mouse. The Table of Contents Wizard enables users to generate links to the chapters of the eBook automatically, based on smart heading extraction. Mobipocket Creator has the flexibility to personalize the cover page.

Mobipocket eBook files have a .prc extension, which is a Palm operating system file extension, but the Mobipocket Reader for both the Palm and the Pocket PC handheld device platforms can read these .prc files. A user merely needs to send the .prc file to her handheld by double-clicking on it, and once the synchronization is complete, the title will appear in the Library of the Mobipocket Reader application. If the Mobipocket Send Book utility does not appear when the Mobipocket eBook is double clicked, it means that the Mobipocket Reader was not installed on the PC. This can be corrected by installing the Mobipocket Reader for the handheld device type onto the computer desktop <www.mobipocket.com/en/DownloadSoft/ProductDetailsReaderPro.asp>. Mobipocket Creator 4.0 currently costs $29.95, but a free preview download is available on their site. Site licenses are also available.

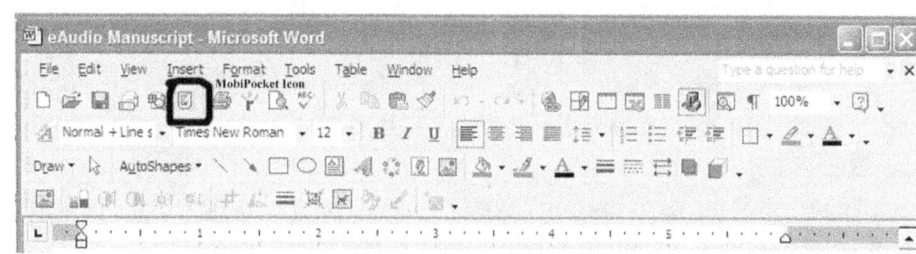

Figure 6:5 Screenshot of MobiPocket Companion Program

Microsoft Word, Excel, PowerPoint, and other Office documents can be converted to the Mobi .prc file format with the Mobipocket Office Companion <www.mobipocket.com/en/DownloadSoft/DownLoadOfficeCompanionStep1.asp>. Mobipocket Office Companion 2.1 is a powerful one-click conversion tool to export any Microsoft Office document to a Palm, Pocket PC, Symbian, or smartphone operating system. All the document formatting, including images, tables, hyperlinks, and table of contents are preserved during the conversion process.

Mobipocket Office Companion 2.0 installs a set of plug-in icons into the Microsoft Office Application suite of Microsoft Word, Microsoft Excel, Microsoft PowerPoint, and Microsoft Outlook®. With the Mobipocket Office Companion Pro version plug-in, icons are also installed into Microsoft Access, Microsoft FrontPage, and Microsoft Visio®. The standard version is $19.95 and the Pro version is $29.95.

WordSmith

WordSmith <www.bluenomad.com/ws/prod_wordsmith_details.html> is a full-featured word processor, document viewer, and enhanced memo pad available for the Palm organizer. WordSmith synchronizes and integrates seamlessly with Microsoft Word so that desktop documents can easily be transferred to handheld devices and vice versa with little or no change in formatting.

Creating eBooks for the iPod

The iPod was designed as an entertainment center, but soon after its release users found it also could be used for educational purposes. The iPod contains tremendous flexibility and can play multiple digital formats, including text. Since the iPod only supports files of 4K size, iPod eBook makers were faced with an additional challenge—cutting the end of a 4K file so that it can automatically find the end of the paragraph, sentence, and word.

To create eBooks in an iPod friendly format, Daniel Duris created an online program at the iPod eBook Creator site <www.ambience.sk/ipod-ebook-creator/ipod-book-notes-text-conversion.php> that lets users upload a plain text eBook file, such as found at Project Gutenberg, and then converts that text file into a notes file for download onto a local computer. This program creates notes for use with an iPod's Notes function. The user merely unzips the downloaded file and then transfers it to the iPod and the eBook is ready for viewing. All notes will be automatically linked, so users can navigate throughout the book with absolute ease.

iStory Creator <www.tucows.com/preview/396253> is an application made to allow people the opportunity to create their very own iPod text games. When certain things happen in the story, the user is able to choose how the story will progress. This feature can lead to games that are similar to a "Choose your own ending" type of novel or it can be developed into quiz-type games, tutorials, demos, references, or anything else with linked text files.

iPod eBook Maker 1.08 <www.freedownloadscenter.com/Utilities/Text_Viewers/Ipod_eBook_Maker.html> is a shareware program developed by Juan Yen that can easily convert .html, and .txt files to eBook files that are fit for iPod, iPod mini, and iPod nano.

iPod eBook Maker 1.20 <http://www.freedownloadscenter.com/Utilities/Text_Form_Editors/Ipod_eBook_Maker.html> or < http://www.softpedia.com/get/Others/E-Book/Ipod-eBook-Maker.shtml> can easily convert .html and .txt files to eBook files fit for iPod, iPod mini, and iPod nano.

Book2Pod <www.tomsci.com/book2pod/> converts text files into iPod notes for the Macintosh.

Summary and Challenge

The possibilities of writing eBooks within the school's curriculum are limited only by the creativity of the teacher and the students. Whatever can be prepared in a text, html, or Word format can usually be transferred to a format that can easily be read by a handheld device. Such sharing and mobility provides the flexibility to individualize student assignments to help them reach their educational goals. Since most students are already interested in the technology, learning to write eBooks with that technology adds an additional dimension and enthusiasm for the learning process. Teachers and librarians can prepare their own eBooks to transfer to student handheld devices. These texts could include study guides or pathfinders that could be easily updated and used from semester to semester.

Before continuing, ask yourself these questions: "What writing projects that your students already do, can be converted to eBook format so students can share them with family and friends?" "What formats would be best for my students' eBooks?" "What study guides and pathfinders that you already do could be converted to eBook format?" "What other materials might you develop for your students?"

chapter seven

Circulating eBooks and eAudiobooks

Whenever new sources of information become available, library media specialists are often the first to consider how those sources can be made available to administrators, faculty, and students. They are discovering that some public libraries have been circulating eBooks and eAudiobooks for some time and are considering if similar techniques might also work in a school environment.

Distributing Applications and Documents

The two basic ways to transfer data in and out of a handheld device is by synchronizing or beaming. Synchronizing can take more than three minutes, so many students learn to synchronize their own handheld devices. Some teachers find it beneficial to assign students to specific computers for synchronizing.

Many teachers prefer beaming documents or applications to other devices. Teachers can only beam to one handheld at a time unless the school purchases a multi-beaming device like WebTarget <http://www.webtarget.com/>. Bluetooth technology allows users to send data to three handhelds at a time. To quickly distribute documents among class members, some teachers beam the application or document to one student and then to another student. Meanwhile, the first student beams it to as many other students as he or she can.

In order to beam several files at once, software such as BeamPro <www.ecamm.com/palm/beampro/> is necessary. If a file is more than 300K, or if users are installing software that requires more than one file to be installed, users should synchronize the files to each handheld. Software packages such as Grant Street's SD Deploy 2.0 <http://grantstreetsoftware.com/Products/sddeploy/index.html> allow teachers, media specialists, and technology coordinators to configure multiple handhelds or install software to multiple handhelds quickly and easily. For approximately $249, the contents of an entire "master" handheld or individual application are copied onto a 32 MB SecureDigital (SD) card that lets users configure palmOne handhelds or Windows Mobile (Pocket PC) handhelds as desired. By inserting the card, SD Deploy copies the configuration to the handheld computers.

Most handhelds do not come with any direct connections to a printer for document distribution, but three options for printing are generally available. The first is by synchronizing the data to the desktop and printing from the desktop. The second is printing directly from the handheld via infrared. Users can also print wirelessly using wireless Ethernet. In order to print with infrared or wireless Ethernet, users need a print utility program such as PrintBoy <www.bachmannsoftware.com>.

Advantages of Circulating Digital Media in School Libraries

Do eBooks and eAudiobooks have a place in school libraries? Should a school library develop a policy addressing the distribution method of these formats? How will eBooks and eAudiobooks help students improve their reading skills or provide sources of information to meet curriculum needs?

Advantages of using eBooks include the ability to change the text's font to suit one's personal taste and size. A tap on the screen will bring up a dictionary definition, pronunciation, and occasionally illustrations. In some cases, the translation of text to other languages is available. Most eBooks will allow annotating, full-text searching, bookmarking, and the adding of notes within the text.

Handheld devices become an ideal "eBackpack" as it stores reference books, magazine and journal articles, notes, teacher-generated study materials, and customized eTextbooks. Not only will the handheld contain all the reference sources, but it can also contain the student's appointment calendar, to-do-list, an address book, a streamed down version of popular software packages, an audio recorder, and a video and audio player.

The price of an eBook reader may be no more expensive than school supplies. Software distributors and eText publishers practically give away eBooks with subscriptions services, and many eBooks are now in the public domain. The funds schools once spent on textbooks and printing costs could be used to subsidize the costs of handheld devices for children whose families cannot afford the one-time investment.

An inexpensive handheld device costs approximately $100 and the average price of a paperback is $4 with an educator's discount. This means that a teacher would break even by choosing to buy a handheld with 25 free eBooks loaded on it.

With a collection of eBooks, media specialists can add thousands of titles to their collection, with all the same rights of ownership as print books, without having to build any additional shelves or storage facilities. Other benefits include no re-shelving and no lost, stolen, or overdue books.

The school's library media center is the logical place to house the infrastructure technologies needed to insure that eBooks connect to each other and the rest of the world. School media specialists are trained to select information in all formats, and make it available to students and staff. They also budget for, acquire, and track the licenses needed to use these products. The media specialist's expertise on the accessibility of digital media may be the single most valuable "resource" the library will offer. They are in the best position to assist teachers in creating individualized student eTextbooks.

Locating eBook Managing Sources

Creating and managing an eBook Library does not need to be as overwhelming as it might appear at the beginning. Several commercial services are available to assist busy librarians and media specialists. The OverDrive® Technology Company <http://overdrive.com/> is a leading provider of enterprise Digital Rights Management (DRM) and content delivery that enables the management and distribution of premium digital content over global networks.

OverDrive's Web-based services provide for secure distribution of eBooks, digital audiobooks, and other digital media throughout multiple channels <http://contentreserve.com/library.asp>. This software allows libraries to build their own eBook and/or eAudiobook collection to meet the diverse needs of their patrons. The patrons can borrow eBooks and eAudiobooks, and access their accounts anytime—day or night—from any computer connected to the Internet. OverDrive is one of the most experienced companies in the world for all of the services associated with eBooks formatted in Adobe Reader (PDF), Mobipocket, and Microsoft Reader, as well as digital audiobooks in Microsoft Windows Media format.

Another content delivery system known as Libwise <www.libwise.com/> allows patrons to download eBooks to either a desktop PC or any of the most popular handheld devices such as Palm, Pocket PC, WinCE, and even some Nokia cell phones. Libwise provides a complete, turnkey eBook lending solution that functions seamlessly with a library's existing OPAC system, and provides a custom Web site branded with the name and logo of the library or organization.

For as low as $29.95 a month, Libwise, hosted and fully supported by Fictionwise, is extremely cost effective, making it practical for small public libraries or school library media centers. Librarians may select content from a growing catalog of thousands of contemporary titles in dozens of categories, both fiction and nonfiction.

Using Libwise, patrons may read borrowed eBooks for the borrowing period configured by the librarian from a few days to a few weeks. At the end of the checkout period, the eBook file expires and becomes available for the next patron to borrow. Libwise allows librarians to choose the model that benefits their patrons the most, such as an unlimited number of patrons may borrow the eBook at a time, or the eBook may be borrowed a maximum of 150 times (simulating "wear and tear" that occurs on physical books).

OCLC Online Computer Library Center, a nonprofit organization that provides computer-based cataloging, reference, resource sharing, and preservation service to libraries worldwide, provides

NetLibrary <www.netlibrary.com> as a content delivery system. Their collection, available for individual libraries, includes more than 100,000 eBooks and hundreds of eAudiobooks and eJournals.

In order to utilize NetLibrary, a library purchases a collection of titles just as they would print titles. NetLibrary then handles all the technology and hosts the eBooks from their own servers. As with print books, only one patron at a time may access each copy of an eBook. The subscribing library sets the checkout time for each eBook through NetLibrary's secure Library Resource Center, which allows librarians to see usage and collection development reports. Local patrons can access eBooks directly through the local library OPAC or via a link on the Web site. Full-level eBook MARC records cataloged in MARC21 format are available for all NetLibrary eBooks.

To check out an eBook through a local library, the patron locates an eBook in the local collection and then has the option of browsing or checking out the eBook. By checking out an eBook, users will have exclusive access to the book during the checkout period set by the library. eBooks are automatically checked back in to the NetLibrary collection until the checkout period expires. At this time, NetLibrary's patent-pending technology ensures that only one person can use one eBook copy at a time. If a library or organization has purchased 10 copies of one title, then only 10 people at a time may view that title.

Patrons can copy or print single pages from an eBook from NetLibrary, just as people can photocopy single pages of a print book. However, NetLibrary has developed mechanisms for limiting the copying and printing of eBooks from the Internet.

NetLibrary is not a subscription service like many other content management systems. The local library owns the eBooks. Librarians can select collections or choose individual titles to create a custom collection. Prices are based on publishers' list prices, with discounts for volume. There is also a service fee, which helps the company manage and upgrade their technology.

NetLibrary subject sets <www.oclc.org/info/k12subjectsets/> provide eBooks for school library or media centers quickly and affordably with eContent solutions that support Web-based research, reference, and learning. Each subject set focuses on a specific type of content, such as biographies, study guides, or science experiments.

Gale Virtual Reference Library <www.galegroup.com/eBooks/> delivers access to reference sources to a subscribing institution's Web OPAC as well as via a customized database collection. With a local library's collection of Thomson Gale eBooks, patrons can search across their entire collection instead of one resource at a time.

Gale Virtual Reference Library allows librarians to customize their reference collection to suit their library's needs and conserves shelf space, while offering unlimited usage. Local libraries can purchase titles one eBook at a time and provide them to their users online with 24-7 access. No specialized reader or hardware is required—just a standard Web browser and the Adobe Acrobat Reader. Gale Virtual Reference Library's searchability technology allows users to search a single eBook or the entire collection. With remote access, library patrons can link to reference content through the local OPAC or navigate directly from the Table of Contents or Index page of a specific eBook.

Gale Virtual Reference Library offers libraries the opportunity to select from an initial collection of approximately 125 reference sources, encyclopedias, almanacs, and series. Each librarian can customize their Gale Virtual Reference Library to fit individual needs. Gale manages the library's collection, showing patrons only the titles that the library has purchased. The selected eBooks can be linked from the library's OPAC through MARC records.

Books24x7 <www.books24x7.com>, a SkillSoft Company, provides online access to the full, unabridged contents of thousands of business and technical books, as well as to expert summaries of popular business books.

Another source of eBooks is a company known as ebrary <www.ebrary.com/corp/libraries.htm>, which provides a research platform that hosts thousands of full-text books, reports, and maps from leading publishers.

Many sites such as Project Gutenberg <www.gutenberg.org/> provide public domain eBooks. With a little ingenuity, a library media specialist or classroom teacher could download eBooks onto an inexpensive handheld device and circulate the device like a print book or magazine.

Creating and Managing an eAudio Library

Several circulation models are in use for circulating eAudio. Some school libraries have chosen to purchase inexpensive MP3 players such as the iPod shuffle™ and check them out with preloaded eAudiobooks. Instead of having an entire book take up several CDs, one book fits within several MP3 files, usually ranging from 150 MB to 350 MB.

eAudiobook collection from NetLibrary <www.netlibrary.com/Librarian/Products/eAudiobooks.aspx> can be downloaded or played on any desktop or laptop device supporting Microsoft Windows Media Player 9 and above, Musicmatch Jukebox v8.2 and above, or Nullsoft WinAmp v5 and above. Library users can transfer their favorite titles to a wide range of portable devices, such as portable music players, portable media centers, Pocket PCs, and even select smartphone devices. NetLibrary and Recorded Books have developed a simplified annual subscription that bases collection price on library size and anticipated circulation requirements.

Once eAudiobooks are available in a library's collection, library patrons can listen to a preview of the audiobooks, and check out and download the audiobooks. Users will need an account with the library to log onto the NetLibrary Web site to be able to access these features. NetLibrary automatically downloads a license with each download, which lets users access the audiobook for 21 days. This license automatically expires after 21 days and the user will not be able to play the audiobook after that, even though the file remains on the person's personal computer and portable listening device. At that point, users will need to delete the file manually from their personal computer and portable listening device.

OverDrive Audiobooks <www.overdrive.com/audiobooks/> with OverDrive technology allows library patrons to select an audiobook and check it out online by accessing the library's Web site. The patron can then use the file for a set loan period. Once downloaded to a PC, the patron will be prohibited from sharing and passing the file to another computer. For portability, the patron will be able to transfer segments of the file, or the entire file, to supported mobile devices, such as handheld computers, MP3 players, or Microsoft smartphones. At the end of the lending period, the file expires and can no longer be opened on the PC by the patron nor repurposed as a portable file.

All audiobooks from OverDrive are formatted for Windows Media Audio. They provide high sound quality in digital stereo, along with many ease-of-use downloading and listening features.

For schools on limited budgets, eAudiobooks can be circulated on inexpensive MP3 players preloaded with public domain eAudio from sites such as LearnOutLoud.com <www.learnoutloud.com/>.

Creating a School eLibrary

Before an eLibrary is created in a school, consider the following situations:

- Where will the eLibrary be located?

- Will the school provide access to the eLibrary vendor's site?

- Will the school support its own eLibrary server?

- Will the school circulate pre-loaded handheld devices?

- Will students be able to access the school network or Internet from computers in the classroom?

- Will the school eLibrary link to eBooks in other libraries or will the eBooks files be located on a local server?

- How will the eBooks be evaluated and selected to match or support the school reading lists and curricula?

- Will the selected books need to be converted to different formats?

- Will librarians and teachers be able to add original student writing to the eBook library?

Summary and Challenge

As school administrators and library media specialists struggle to improve library services with shrinking budgets, more and more are turning to circulating eBooks and eAudiobooks to meet the needs of their students and faculty. By using services already available, schools now can make digital books available to their students that would not be available in print or CD format. Portable devices that were once considered strictly a means of entertainment can easily be turned into devices for education.

Before continuing, ask yourself these questions: "How would an eLibrary enrich your school's curriculum and lead to higher test scores?" "Would it be possible to set up an eLibrary in your school? If so, what formats would you use?" What methods of distribution would work the best in your environment?"

chapter eight

Utilizing and Preparing eAudio

Some educators are reluctant to use audiobooks for fear it might replace a child's desire and motivation to learn to read. However, most experts maintain just the opposite is true. Audiobooks are a powerful tool that allows students to read above their actual reading level. Listening to audiobooks is especially significant for older students who may still read at a beginner level. While these students still require time to practice reading at their level, they must also have the opportunity to experience the characterizations, plot structures, themes, and vocabulary of the same books their friends are enjoying.

Educational Value of Audiobooks

Just as reading literacy and visual literacy are vital to a well-rounded student, so is audio literacy. Listening is an extremely complex, interactive skill by which spoken language is converted into meaningful information within the mind. The use of audiobooks helps students meet Information Literacy Standards for Student Learning (*Information Power*) as students develop higher-level thinking skills and learn to draw conclusions and develop new understandings as they respond to literature in audio format.

By listening critically and responding to the narrator's interpretation of the author's words, students can distinguish the point of view of the book through the cadence and tone of the reader. Audiobooks also model fluid phrasing and cadence, assisting with comprehension based on a fluent, not word-by-word, reading. The understanding of correct pronunciation of English, dialects, and non-English words is enhanced in multicultural audiobooks. Teachers can inadvertently convey a lack of confidence in reading these books aloud when they are unfamiliar with words in a foreign language or can be reluctant to read aloud the books in which the rhythm of the text feels unfamiliar when spoken. When paired with matching text, audiobooks reinforce words, phonics, and syntactic knowledge to assist a reluctant reader's transition to fluency.

The mechanics of listening to an audiobook are different from listening to music. When the listener turns off the audio player and then turns it on again, the audiobook begins playing exactly where it had left off. In addition, audiobooks are divided into chapter markers so listeners can easily navigate to different parts of the book.

Besides the educational value of audiobooks, they also provide recreational reading/listening experiences for students based on individual needs and interests. Audiobooks provide students a motivating format, promote reading enrichment, and encourage enjoyment of creative expressions of information. Many readers will reach for an audiobook as a way to re-experience something they have enjoyed in print. While listening, previous readers may notice the humor they missed in their rush to find out what happened next, or they may fall in love with the descriptive prose that their eyes skipped in the reading. The art of the audiobook is to provide a genuinely different sensory experience.

The International Listening Association has compiled interesting Listening Factoids at <www.listen.org/pages/factoids.html>. They maintain that 85 percent of what we learn, we learn by listening; 45 percent of the average day is spent listening; and most listeners normally retain about 20 percent of content. They also maintain that listening can be improved through practice and instruction. For these reasons, the use of eAudiobooks needs to be encouraged among both students and adults.

Most educators recognize the value of listening skills (e.g., main ideas, details, sequencing, predicting) as well as processing auditory information, but few recognize the impact audiobooks can have on improving listening skill and expanding a student's attention span.

The average person, whether student or adult, spends several hours a week when their eyes are busy, but their mind may not be. This is especially true when commuting to and from school and community events. Instead of this time becoming wasted time, it would be more advantageous to listen to audiobooks or articles. The availability of the proper hardware, software, and resources can turn wasted moments into intellectually stimulating moments. Words become the new music for the audio player.

Selecting Audiobooks

Audiobook listeners and educators may feel overwhelmed with the number of audiobooks available to them. To meet this need, The American Library Association prepares an annual list of Notable Children's Recordings Awards <www.ala.org/ala/alsc/awardsscholarships/childrensnotable/notablecreclist/currentnotable.htm> and a list of Selected Audiobooks for Young Adults <www.ala.org/ala/yalsa/booklistsawards/selectedaudio/selectedaudiobooks.htm> to assist librarians, students, and parents with the selection of quality audio recordings.

Audiobook Industry

The Simply Audiobooks Web site <www.simplyaudiobooks.com/processInterfaceAction.php?pId=138&rId=3> provides a fact sheet as to the state of the Audiobook industry. Relevant facts include:

- "In 2004, the Audio Publishers Association (APA) estimated the size of the audiobook market is $800 million, factoring in sales from non-reporting and non-APA members.

- In December 2004, audiobook publishers reported a 14 percent increase in retail and wholesale sales between 2002 and 2003. – *APA*

- The sector showing the most growth in the audiobook industry is in downloadable sales. Download of audiobooks increased 50 percent year over year from 2001 to 2003, with a six percent increase in 2005. – *APA*

- In libraries, circulation for adult audiobooks jumped 13.5 percent in two years, while circulation of children's audiobooks rose 10.7 percent. Budgets for adult audiobooks increased 6.1 percent and those for kids audiobooks increased 4.8 percent. – *APA*"

AudioFile <www.audiofilemagazine.com/> is the only magazine devoted to audiobooks and is indispensable for anyone who enjoys and utilizes spoken-word audio. The magazine's focus is the audio presentation, not the critique of the written material. The printed guide can be ordered from their Web site and is available to premium subscribers online.

AudioFile publishes a comprehensive reference guide to audiobook titles. Its editors compile the year's 80 best audiobooks for children and teens, recognized on notable lists by librarians and teachers from the children's and young adult divisions of the American Library Association to Capitol Choices.

After a user loads a number of audio programs on a desktop, laptop, or handheld device, it can become confusing to remember and locate those programs. To help organize the audio loaded on a handheld device, an outside program is helpful. AudioList™ <www.wakefieldsoft.com/audiolist/> is a complete audio, CD, MP3, and music organization and inventory software for Palm OS handhelds and Pocket PCs. AudioList is a database for importing MP3 file libraries and includes the ability to download CD information from the Internet for automatic entry. With this software, users can use multiple sorts, filter options, and define their own genres, ratings, labels, and status lists.

Sources of Audiobooks

LearnOutLoud.com <www.learnoutloud.com/> is believed to be the first and most comprehensive database of audio and video learning publishers and retailers on the Internet. This directory features free audiobooks, lectures, speeches, sermons, interviews, and many other free audio and video resources. Most audio titles can be downloaded in digital formats such as MP3 and most video titles are available to stream online. LearnOutLoud.com Free Audio <www.learnoutloud.com/Free-Audio> provides a wealth of information to meet audio and video learning needs.

Audible.com® <www.audible.com> is the Internet's leading provider of spoken audio entertainment, information, and educational programming. Audible has 80,000 hours of audio programs from 270 content partners that include leading audiobook publishers, broadcasters, entertainers, magazine and newspaper publishers, and business information providers. The Audible.com file format provides book marking and chapter stops, and a state-recall that remembers where the listener last stopped.

Audible.com is also the main provider of spoken audio for Apple iTunes' Music Store and provides digital distribution of spoken content for Random House, Inc. Audible Education provides resources for students, teachers, parents, higher education faculty, and professionals.

Other sources of audiobooks in MP3 format include Audiobooks.com® <www.audiobooks.com/>, which claims to have the world's largest selection of new and used audiobooks on cassette and compact disc, and now provides an electronically downloadable format from PayPerListen.com.

Simply Audiobooks <www.simplyaudiobooks.com/> provides rental, download, and purchasing options.

AudioBooksForFree <www.audiobooksforfree.com/> specializes in audio versions of older books whose copyrights have expired. As the name implies, AudioBooksForFree.com offers its highest compression format for no cost. Most of the free titles, except children's books, include advertisements. Higher quality versions of the same titles are available for a fee.

Fictionwise <www.fictionwise.com/>, a popular source for eBooks, now contains almost 300 downloadable Audiobooks in MP3 format at <www.fictionwise.com/Z130748H9335/ebooks/audio.htm>.

LibriVox was started in August 2005 as a volunteer project. LibriVox <http://librivox.org/> provides free audiobooks from the public domain. Most of their texts are from Project Gutenberg and the Internet Archive. Volunteers record chapters of books in the public domain, and then the audio files are released back onto the Net as a podcast or in a catalog. The objective of LibriVox is to make all books in the public domain available, for free, in audio format on the Internet.

Soundsgood.com™ <www.soundsgood.com/> provides members access to thousands of titles. Their audiobook titles are built using the Microsoft Windows Media audio format, which greatly enhances the quality of the audiobook but can take more storage space than MP3 formats.

A site called American Rhetoric <www.americanrhetoric.com/top100speechesall.html> provides MP3 versions of some of the greatest speeches from recent history. Additional audio learning opportunities include college lectures offered through The Teaching Company <www.teach12.com/teach12.asp?ai=16281> and Modern Scholar <www.recordedbooks.com/index.cfm?fuseaction=scholar.home>.

PlaylistMag.com <www.playlistmag.com> is the ultimate iPod guide online. The availability of audiobooks at the iTunes Music Store is a tremendous improvement to their collection, but their store does not have as wide a selection as Audible.com. In some cases, the same material is less expensive on Audible.com than iTunes. By using iTunes 3 and the iPod Software 1.2 Updater from Audible.com, users can play Audible.com material on their iPod.

Recorded Books <www.recordedbooks.com/> is a premier educational and professional publisher dedicated to creating books, audio products, periodicals, software, and online services for the K-12 supplemental education, public and school library publishing, and audiobooks markets. Recorded Books helps put every student on the same page in an inclusion classroom by helping a pullout group get through a difficult text and allowing a struggling reader the opportunity to read independently by providing a student with access to a great book. School users can now download unabridged audio directly into a school library or classroom with one subscription.

Ripping Audiobooks

If a school already owns audiobooks on CD, it is a simple matter to change their audio format (commonly known as ripping) so they can be used on favorite audio players. Most traditional music rippers will also work with spoken audio. However, care needs to be taken that copyright laws on individual titles will not be violated in converting formats.

In converting audio formats, it is easier to manage audiobooks if they are ripped as a few large files instead of lots of small ones so that listeners do not lose a chapter along the way. Because audiobooks do not need the sound fidelity music does, they can be encoded at a lower bit rate. Digital audiobooks generally use a 32-Kbps bit rate. The advantage to using a low bit rate is that files will be much smaller, so long books can be held on a small-capacity MP3 player.

Kirk McElhearn states in the article "Ripping Audiobooks" in the online version of *PC Magazine* <www.pcmag.com/article2/0,1895,1846814,00.asp> that iTunes is one of the best programs for ripping audiobooks because it permits users to rip CDs at low bit rates. Windows Media Player will not let users rip MP3s at less than 128 Kbps (kilobits per second). To rip an audiobook in iTunes, users must open the "Preferences" and choose a format (AAC for iPod, MP3 for other players), select "Custom" from the Settings menu, select the 64-bit rate, and then close the preferences.

ID3 tags allow MP3 players to display the track, album, and artist information. With the audio editor Audacity® <http://audacity.sourceforge.net/>, users can add tags when they export their file to their MP3 player, or they can add ID3 tags in iTunes, by selecting the recording and choosing "View Info."

Direct Audio on a Handheld Device

Voice recorders have long been a tool for college students to avoid getting writer's cramp while taking class notes. Now, with the increase of handheld devices and MP3 Players, students, teachers, and administrators are able to use the same device for a multitude of tasks, including audio recording.

Many handheld devices come with built-in microphones and voice recorder features. If voice-recording features are not included in the original configuration of a device, add-on accessories may be purchased separately. Griffin iTalk <http://www.griffintechnology.com/products/> turns an iPod into a digital voice recorder for directly storing and playing back audio.

When users plug an iTalk into the iPod's headphone and remote outputs, the iPod goes straight into voice-recording mode.

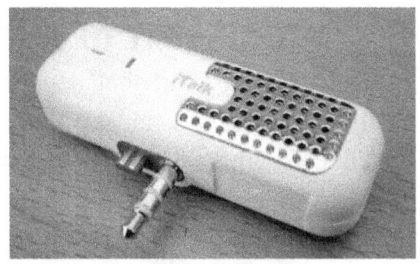

Figure 8:1 iTalk Standard (top) and Pro (bottom)

In this mode, the iPod displays a large, stopwatch-like counter that marks recording time in hours, minutes, and seconds. Two menu choices below the counter offer record and cancel options. The iTalk saves recordings by date and instantly transfers them to a computer when the iPod is synced. The iTalk also contains a built-in speaker, letting users review recordings or play back recordings through its headphones connected to a pass-through headphone jack. This feature allows users to monitor their voice recording or simply listen to the music without having to remove iTalk.

Another voice recording option is the Belkin® iPod Voice Recorder <http://catalog.belkin.com/IWCatProductPage.process?Product_Id=158384> that contains an omni-directional microphone that records in mono sound at 8 KHz and 128 Kbps, or about 1MB per minute of audio. The iPod Voice Recorder records audio as WAV files without any options for recording in other formats. Although the Griffin Technology iTalk and the Belkin iPod Voice Recorder's interface, setup, mike level, and file format are the same, a review in the August 18, 2004, online issue of *MacWorld* <www.MACworld.com/2004/08/reviews/ipodvoicerecorders/> states that the iTalk "is far superior. It's less expensive ($40), it has more features, and it does a better job overall than the Belkin."

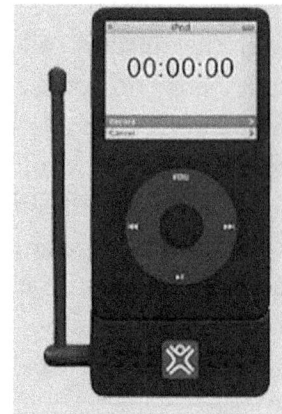

Figure 8:2 MicroMemo

One of the highly recommended digital voice recorders is the MicroMemo™ <www.xtremeMAC.com/audio/earphones_recorders/micromemo.php>. The MicroMemo sells for $79 and is compatible with the iPod Video. This digital voice recorder uses iPod high-fidelity recording (44 KHz/16 Bit - high quality recording). It comes with a removable microphone with flexible neck and is compatible with any 3.5mm-plug microphone. It has no internal battery but receives its power from the iPod itself.

The Audacity product family targets doctors, lawyers, law enforcement, and other professionals requiring professional-grade dictation. Each Audacity product functions as a stand-alone item, but integrated together, provides a seamless solution. Audacity Digital Voice Recorder <www.audiost.com/transys.htm> is a software product that works on many handheld devices that include voice-supported hardware based on an intelligent voice recorder. Audacity also includes programs that integrate the dictation process with database information and forms.

Converting Text to Audiobooks

Educators often find that a particular eBook would be extremely beneficial for a student, but the reading level is too far above that student's ability for it to be beneficial for her. To meet this need, an educator may turn to a text-to-speech (TTS) conversion program. Text-to-speech software can convert documents into spoken words. These programs help students read their favorite electronic archive, storybook, Web page, or rich text document. Instead of taking hours to read a book or article into a microphone, text-to-speech software provides a method to prepare audiobooks within minutes.

Improvements have been made in recent years to remove the mechanical, monotone sound of the speaker in these programs, and they are now more natural to the ear. However, before converting an eBook to an audiobook, the copyright status of the book should be consulted.

TextAloud 2.0 <www.nextup.com/> uses voice synthesis to convert text into spoken audio in MP3 or WMA formats. Users can read text from email, Web pages, reports, and books by listening on their PC, using portable devices like iPods, Pocket PCs, or CD players.

Microsoft Text-to-Speech Package <www.microsoft.com/reader/developers/downloads/tts.asp> can enhance eReading with new accessibility options for Tablet PC and Microsoft Reader for Windows-based PCs and laptops. It is available in English, French, and German. Microsoft's documentation recommends that TTS be used only for phrases or short text passages.

Read Out Loud is a TTS tool that is built into Adobe Reader 6.0. <www.adobe.com/enterprise/accessibility/reader6/sec2.html>. This feature reads text contained within a document window.

Learning Foreign Languages

VITO SoundExplorer <http://vitotechnology.com/en/> is an audio player and audio recorder for Pocket PC that offers a comprehensive set of tools for educational and hobby use. This software allows users to listen to any audiobook in VITO SoundExplorer while they read the text as an eBook. This helps users learn to speak foreign words accurately as they sound in the native language.

VITO SoundExplorer allows users to slow down the playback in order to make out every word the dictator is saying without any quality loss. This program also allows users to record themselves speaking the foreign language to check their pronunciation. VITO SoundExplorer offers CD quality MP3 recording so users can hear their voice as it sounds in reality.

For the Disabled

Listening to eAudiobooks is extremely helpful for the visually impaired. Students with vision limitations are encouraged to enroll in Recording for the Blind and Dyslexic® (RFB&D) <www.rfbd.org/> since many textbooks are available in audio format. RFB&D is a nonprofit volunteer organization and the nation's educational library serving people who cannot effectively read standard print because of visual impairment, dyslexia, or other physical disability. Other audio resources for the visually impaired can be obtained through most state departments of education.

The Mid-Illinois Talking Book Center <www.mitbc.org/> provides free library service for anyone unable to read regular print because of a visual or physical disability. They have books and magazines in audio format along with playback equipment. Books and magazines are mailed free to and from library patrons, wherever they reside. There is no charge to the patron.

For those with dexterity limitations, VITO Voice2Go <http://vitotechnology.com/en/products/voice2go.html> is a voice recognition application for the Pocket PC that allows managing a handheld device with one's voice. With VITO Voice2Go, users no longer need to tap the screen with the stylus or learn a specialized voice recognition program. VITO Voice2Go is able to start and quit applications, call contacts and hang up the phone, modify system settings, or press any buttons with the voice and macro recording.

Online Radio

With the appropriate software, with Wi-Fi connections, many handheld devices can be adopted to be a traditional radio and stream Internet radio. Radio4PDA <www.radio4pda.com/> is a mobile portal that lists streaming radio stations in the United States, United Kingdom, and Ireland. If users have a wireless connection and like listening to the radio, this site will help them find the station they desire. The United States section is organized by state, letting users find their local stations.

MSMobiles.com <http://mobile.msmobiles.com/radio.php> has a mobile site with links to some of the most popular streaming sites in English. With devices with Wi-Fi capabilities, users need to click only on the stream they want to listen to, and Windows Media Player will launch and begin playing. Links include CNN Radio, NPR – National Public Radio, Bloomberg Radio, Air America Radio, the BBC, and other popular stations.

Resco Pocket Radio <www.resco.net/pocketpc/radio/> is a player for Internet streaming radio broadcasting in MP3 or Ogg Vorbis format. Resco Radio includes a vast directory of Internet radio stations. This program not only lets users add categories and radio stations, but it also can import M3U or PLS files that can be downloaded from radio station Web sites.

PocketGear.com <www.pocketgear.com> provides several satellite or Internet software packages developed by eBook Software Company. Pocket XM Radio Satellite Radio 1.1 is the first to merge XM radio, America's number 1 satellite radio, with Pocket PC. PocketStreamer Pro energizes a Pocket PC with LIVE streaming TV, radio, music, news, and weather. Pocket PC needs to have Internet access and Windows Media Player.

Summary and Challenge

Instead of using multiple devices for education and entertainment, a single handheld device can be adapted for a number of purposes and eliminate the number of gadgets and wires needed. The availability of free audiobooks for downloading is revolutionizing the manner in which schools can provide information to students.

With the increased need to mainstream students with Individual Educational Plans, handheld devices can help them access the same information and remain with their classmates. To achieve this, text can be converted easily to audio and uploaded to a handheld device for the student to listen and review at his convenience.

Students, teachers, and librarians are now able to better utilize travel time by using eAudio with educational value. eAudio appeals to the multiple intelligence of individuals. While the use of reading should never be minimized, auditory learners can gain tremendously by having informational audio within easy access.

Before continuing, ask yourself these questions: "How might audio improve your students' test scores?" "Where might you locate that audiobook?" "Will it be helpful to convert text to audio for students in your class with limited reading skills?" "How might the library media specialist assist in providing audiobooks for your students?"

chapter nine

Utilizing and Preparing Podcasts

What Is Podcasting?

Podcasting is an exciting new audio medium that gives educators and students the freedom to listen to the audio program of their choice at the time of their choice. Podcasting is a new word created by joining two words and their meanings—iPod and Broadcasting. The difference between a radio broadcast and a podcast is that the radio broadcast requires a studio and a transmitter that sends out radio signals within a given area. That signal can be received by anyone who happens to be in the area with a radio tuned to the appropriate frequency. Podcasts, on the other hand, are audio programs that are stored as digital files on the Internet. These files can be downloaded from an Internet connection anywhere in the world.

Neither producing a podcast nor listening to one requires an iPod. The name came about simply because Apple Computer's iPod was the best-selling portable digital audio player when podcasting began. The power of podcasting is its ability to deliver content to any MP3 player, laptop, or desktop computer that is capable of synchronizing with Windows Media Player, Apple iTunes, or RealPlayer.

Podcasting is a time-shifting technology. Learners can download an audio file to a portable device and listen to it at their convenience. For some students, this might mean they would listen to a podcast in the quiet of the library or at home, while others might take advantage of downtime while commuting, walking across campus, or waiting in line.

The podcast is a descendant of Web logs, or blogs. The main differences between the two are the formats and delivery methods of the information. Blogs can be memos, articles, personal reviews, journals, or other forms of written content delivered on the Web through a browser. The text in a blog may be accompanied by links and images, which could include links to video and audio files.

Podcasts are pre-recorded audio files that can be posted to text-based blogs in the form of MP3 music files. Podcasting is a process of automatically receiving time-shifted audio or video from a personally selected subscription via a podcast-enabled RSS 2.0 feed down to a portable media device through the Internet. The podcast users are able to control what content they listen or view, and when and where they utilize it.

Subscribing to podcasts allows a user to collect programs from a variety of sources for listening or viewing offline at whatever time and place she chooses. In contrast, traditional broadcasting provides only one program at a time specified by the broadcaster. MP3 music and speech files have been available over the Internet for several years, but podcasting makes it simple for individuals to record and upload their own programs.

Users have been able to post MP3 music and speech files to a Web site for download for several years, but today they can set up a podcast for others to subscribe to so they will get the MP3's automatically. Podcasts are possible because of RSS, or "Really Simple Syndication," technology that is used to send text news feeds to Web sites. RSS feeds are written in extensible markup language (XML), which can be imported and displayed by a variety of different types of applications. Podcatcher software imports RSS data to manage the downloading of podcasts. Whenever a user wants to subscribe to a Web site's RSS feed, all they have to do is copy that URL and paste that information into the podcast-supported client's Add Feed window.

The ability to "aggregate" programs from multiple sources is a major part of the attraction of podcast-listening. Also, podcast producers have more flexibility because they do not have to make advertisers happy and they do not have to worry about FCC regulations or pay attention to the corporate bottom line.

In a short time, podcasting has come a long way and quickly obtained the backing of educators at the elementary, secondary, and high education level as a tool for the active, mobile learner. Podcasting involves a shift from e-learning to m-learning. E-learning, or electronic learning, refers to any computer-based learning that enables students to access and make use of course materials at a distance and at their convenience. M-learning, or mobile learning, capitalizes on the increasing ubiquity of wireless networks and devices such as laptops, handheld computers, wireless phones, MP3 players, and of course, iPods.

The reason for the overnight success of the podcast is attributed to the millions of portable MP3 players being used, with gigabytes of empty storage space and the MP3 format itself, which is a part of all these portable players.

History of Podcasting

Former MTV video jockey Adam Curry and Dave Winer, the developer of the RSS 2.0 specification, are generally considered the "podfathers" of the podcasting community. In October of 2003, Winer and his friends organized the first BloggerCon conference at Harvard Law School's Berkman Center for the Internet and Society in Cambridge, Massachusetts. After much work, by September, 2004, Adam Curry catalyzed the podcasting concept when he released a script that automatically downloaded audio files referenced in RSS files. The word about podcasting rapidly spread through the already-popular Weblogs of Curry, Winer, and other early podcasters and podcast-listeners.

On October 11, 2004, the first phonetic search engine for podcasting, called Podkey, was launched to assist podcasters in connecting with each other. In March 2005, vice-presidential candidate John Edwards became the first national-level U.S. politician to hold his own podcast. Then in May of that same year, the first book on podcasting, *Podcasting: The Do It Yourself Guide* by Todd Cochrane, was released. In June 2005, Apple announced support for podcasts in its iTunes software, with distribution through its iTunes Music Store. Within two days, customers had subscribed to more than one million podcasts from Apple's then-available 3,000 selections.

Podcasting, unlike blogging, has been instantly, if cautiously, recognized by big media for its potential, and most nationwide news programs are now available in podcast format. In some podcasts, paid advertisements are beginning to slip in. Mobilcast™ from Melodeo <http://mobilcast.com/> is one of the first attempts to provide podcasting to cell phones. As soon as the price of cell phones with multi-GB hard drives comes down, the distinction between cell phone and iPod will become blurred.

Educational Uses of Podcasts

When teachers are provided the necessary equipment, software, and resources to prepare a podcast, they soon develop ways to utilize podcasting to improve reading fluency and writing skills. Not only do students listen to educational podcasts, but they also develop their own. In this preparation of their own podcasts, they learn to research, plan, and write the scripts in advance and complete multiple audio takes before they're satisfied that the quality is good enough for broadcasting.

In the Fall of 2005, Stanford University began a partnership with Apple iTunes Store to publish and host lectures called *Stanford on iTunes*. iTunes U. (for University) became a partnership between Apple and individual colleges and Universities for hosting and distributing audio and video lectures, podcasts, and vodcasts to their student bodies. Modeled after the Stanford on iTunes program, iTunes U. is a free service and allows a school to create an environment for instructors to upload their audio and video podcasts for distribution to their student bodies.

Colleges and universities are marketing the benefits of their programs to high school students or potential transfers considering the university through podcasts. Some potential students may be able to listen to several class sessions in their area of interest, a message from school officials, coaches, or students themselves.

However, the interest in educational podcasting is not limited to the secondary or higher education level; elementary teachers are also utilizing podcasting as an instructional tool. Tony Vincent, a fifth grade teacher and technology specialist in Omaha, Nebraska, pioneered the use of handheld computers in elementary school. When podcasting was in its infancy, he pioneered the use of podcasts within the curriculum <http://www.learninginhand.com/podcasting/index.html>. He has recently collaborated with Mike Curtis to produce a bi-weekly podcast for educators called Soft Reset <http://www.learninginhand.com/softreset/>. Listeners will learn helpful tips for using handhelds in education, where to find useful resources, and listen to insightful discussions on the topic.

Dave Warlick, author of *Redefining Literacy for the 21st Century* by Linworth Publishing, and owner and principle consultant for the Landmark Project, a Web site devoted to using technology in education in practical ways, has recently created The Education Podcast Network <http://epnweb.org>.

This network is an effort to bring together the wide range of podcast programming that may be helpful to teachers looking for content to teach with and about, and to explore issues of teaching and learning in the 21st century. Most of the producers of the programs listed on the EPN network are educators who have found an opportunity where they can share their knowledge, insights, and passions for teaching and learning.

Another directory of podcasts for educational use and suitable for use by children and young people at school, college, and elsewhere is located at <http://www.recap.ltd.uk/podcasting/channels/podchannels.php>. This directory developed in the United Kingdom covers podcasts from around the world that are produced and published by pupils, young people, and educators.

For several years, K12 Handhelds, Inc. <http://www.k12handhelds.com/podcasting.php> has led the way in providing schools with integrated solutions for mobile technology use in K-12 education. Soon after podcasting was developed, they provided guidelines and resources for using and producing podcasting in the K-12 curriculum. Another resource on podcasting is a book for educators called *KidCast: Podcasting in the Classroom* by Dan Schmit, an instructional technologist in the College of Education and Human Sciences at the University of Nebraska-Lincoln.

Not only can podcasts be used as a curriculum tool, but they can also become a communication tool for library media specialists, teachers, and administrators. RSS technology can be incorporated into the school system Web site or a school Web site. With podcasts, administrators can deliver audio recordings of daily news, press announcements, board meetings, or even weather-related closings. Teachers can use daily or weekly podcasts to summarize lessons for students and parents.

By using podcasts, teachers are able to personalize instruction to facilitate self-paced learning for remediation of slower learners, to offer advanced or highly motivated learners extra content, or to provide audio experience for multilingual education. Podcasts also provide the ability for educators to feature guest speakers from remote locations as well as allowing guest speakers to present once to many sections of a class.

More schools are turning to recording class sessions and converting them into podcasts. Now when students miss classes they no longer have to ask classmates to fill them in on what they missed. Instead, they could download audio files to their MP3 players or personal computers. Teachers can even use their cell phones to create a podcast with daily homework assignments and other classroom information. Parents can then download the podcasts to stay up-to-date on their children's school activities.

Commercial vendors have begun to include podcast feeds as part of the database subscriptions they offer to schools. Thomson Gale has announced the addition of podcast feeds to *InfoTrac* on Thomson Gale *PowerSearch* and its *Student Resource Center*, *Opposing Viewpoints Resource Center* and *History Resource Center* databases. <http://pressroom.gale.com/index.php/2005/11/thomson-gale-adds-podcast-feeds-to-database-resources/>. Weekly presidential radio addresses began in January 2006 in podcast format.

Educators have formed numerous online groups to exchange podcasting ideas. One of the most popular is a Yahoo Group listserv, Podcasting in Education. <http://groups.yahoo.com/group/Podcasting-Education/>.

Examples and uses of student and teacher produced podcasts that support the K-12 curriculum includes the following:

- Grade 5 Audio Poetry Journal   <http://blue1.emerson.u98.k12.me.us/magazine/audiopoetry05>

- Bob Sprankle's combined third and fourth grade class in Wells, Maine publishes a podcast at Room 208 <http://bobsprankle.com/blog/>. Room 208's Student Updates involve groups of four students in writing a segment about learning over the course of the week.

LearnOutLoud.com <http://www.learnoutloud.com/> contains a Podcast Directory of free educational content that is selected to instruct, inspire, and enlighten the listener. This directory includes links to podcasts of tutorials on learning a foreign language to listen to classical literature.

Legal and Copyright Issues in Podcasting

Unlike other forms of radio broadcasting, podcasting as it is broadcast today is not subject to direct censorship or regulatory control. This is part of its attraction. However, there are pitfalls for the naïve and unsuspecting. Just like any other form of published media, content on the Internet is subject to copyright, associated terms of use, and other laws such as defamation.

The laws covering copyright, offensive material, and libel are complex. Schools, like individuals, must take steps to ensure that published podcasts do not infringe on any existing copyright and that the content is not offensive or libelous. Teachers can turn these challenges into a teachable moment with a discussion on citizenship and individual respect and responsibility.

One of the problems of the non-regulated podcasts is that there are no controls against explicit content. Parents and educators need to engage young people in conversation about what they are listening to, what they think about it, and how it applies (or does not apply) to their schoolwork or the development of their character.

Regardless of local school policies, educators should always request permission and inform parents when students will be publishing their work online, whether it is text, photo, or a podcast. Parents should be assured that care is taken to protect their child's identity and that they will be supervised while online.

Portable digital recorders allow podcasters to make their recordings anywhere. It would be easy to record conversations without all the participants being aware that they are being taped. It is also possible to use software to record online conversations in audio chat programs such as Skype <www.skype.com/>, iChat <http://www.apple.com/macosx/features/ichat/>, or Gizmo <www.gizmoproject.com/>. In most cases, it is unethical to secretly record conversations.

The use and exchange of music over the Internet has been one of the most complex issues educators face. To use music in a podcast, students and teachers have three options.

1. Arrange for music licenses that cover performance rights and royalties.

2. Use royalty free music (carefully check the conditions of use).

3. Create their own, original music (this is their own copyright).

In seeking permission to use music that is protected under copyright laws, one first needs to consider who is the actual copyright owner of a musical work and how a license can be obtained to transmit a musical composition on the Internet. Generally, the songwriter and/or his or her music publisher are the owners of the rights to a musical composition. Licenses to perform a musical composition can be obtained from organizations that represent a group of songwriters and publishers.

ASCAP (The American Society of Composers, Authors, and Publishers) <http://ascap.com/weblicense/> has worked with many operators of Internet sites and services to develop one of the best licensing solutions for the expanding number of online music uses and business models. Currently ASCAP provides two new versions of their widely used Internet license agreements: "Non-Interactive 5.0" <http://ascap.com/weblicense/release5.0.pdf> for non-interactive sites and services, and "Interactive 2.0" <http://ascap.com/weblicense/release2.0.pdf> for interactive sites and services.

A popular Internet site to help users obtain copyright permission on specific works is The Harry Fox Agency, Inc. <www.harryfox.com/> that provides a musical copyright information source and licensing agency. Schools, churches, community groups, and other low-volume producers of audio can quickly and easily obtain a license online at <http://www.harryfox.com/public/songfile.jsp>.

As an alternative to playing music controlled by the RIAA (Recording Industry Association of America), podcasters can promote musicians who are not signed to a label and are not members of the RIAA. Some music can legally be played under a Creative Commons License <http://creativecommons.org/>. Creative Commons makes available licenses and tools to enable creators and licensors to license their works on more flexible terms. The Creative Commons License requires crediting the artists and not pursuing commercial uses. Some of the music licensed through Creative Commons is found on GarageBand <http://GarageBand.com>. However, not all GarageBand.com artists have a Creative Commons License, and permission must be requested directly.

Royalty free podcast music is available from numerous online sites. Such sites include the PodSafe Music Network <http://music.podshow.com/> and Podsafe Audio <www.podsafeaudio.com>. Both let independent musicians post free music podcasts and users must adhere to the Creative Commons License, which requires crediting the artists and not using the music for commercial gains.

The creator of a podcast is the copyright owner of that podcast. The words that students speak belong to them. When they are published as a work, they are protected under copyright laws. By placing a copyright symbol and date on their work, or copyright element information within the RSS feed or MP3 file's ID3 tags, work can be properly attributed to the owner of the intellectual property. Just as students respect the copyright laws of the music producers, so also the students' work should be respected for their copyright ownership.

Hardware Needed to Listen and Produce a Podcast

Any computer purchased within the last two or three years with 512 MB of RAM, three to four GB of free space on the hard drive, Line Out/Line In jacks, and capable of running Windows XP/2000 with a Pentium 4 or a Macintosh running OSX, should be capable of creating and receiving a podcast.

A full featured USB or FireWire audio to digital converter provides more flexibility and higher quality sound. These devices work with higher quality XLR microphones, and most will support two simultaneous microphones, making it easy to record a two-person interview. These also come with earphones to monitor the recording.

Some educators may choose a complete podcast hardware package such as Podcast Factory <www.m-audio.com/products/en_us/PodcastFactory-main.html> priced at approximately $180 that works with either Windows XP or Mac® OS X 10.2.8 and later versions. The Podcast Factory combines all the hardware and software needed to record, edit, and publish professional-sounding podcasts. It even includes software that processes MP3 files and automates Web publishing of RSS 2.0 feeds. This package includes a USB-based audio interface box that adds high-quality ports to connect a microphone, headphones, and musical instruments to the computer, plus a microphone with a desk stand and software to edit the recordings for publication on the Web.

Microphone quality can range from a $10 microphone from Radio Shack to a professional grade one. Microphone upgrades are the number-one purchase that will improve the quality of a podcast. Types of microphones include the common dynamic microphones, which usually require no power source, and the condenser microphone, which requires phantom power (external power). If a microphone is chosen that requires phantom power, users will want to make sure that the mixer they buy has phantom power.

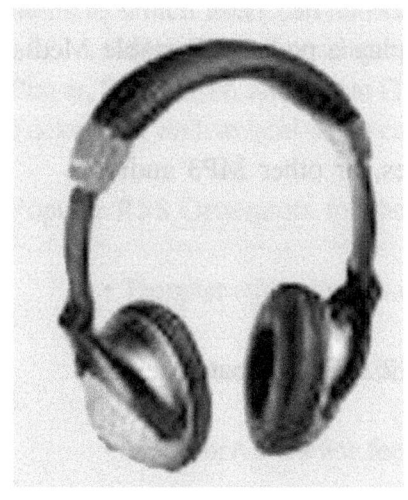

Figure 9:1 Full-cuff Headset

Any full-cuff headset will enable podcasters to hear the noise that is being introduced into the recordings and get a real sense of what is being recorded. A full line of headsets and headset reviews can be found at <www.headphones.com/>.

There is a wide difference in quality and price of audio mixers. The basic mixer needed by a podcaster starts at $49.95 but can top out as high as $500. Examples of possible mixers can be found at <www.tapcoworld.com/products/index.html>.

While basic podcasts can be produced without a mixer, the use of a mixer will provide these advantages:

Figure 8:2 Sound Mixer

- Ease of manually controlling volume on background music via the mixer while talking rather than dealing with a mouse and trying to move a slide bar on a media player.

- High quality, low-noise microphone amplifiers.

- The capability to insert external audio and multiple microphone inputs.

- The ability to record telephone interviews.

- The capability to patch in special effects.

If telephone interview equipment is desired, a simple telephone line interface from Radio Shack <http://radioshack.com> known as Smart Phone Recorder Control, which costs less than $30, will work. For more features, JK Audio <http://jkaudio.com/inline-patch.htm> provides an Inline Patch for $270 that can be used with talk shows when users may need access to audio from a working telephone. The

For people who use several different computers throughout the day, a podcatcher has been developed so podcast listeners can use any computer with an Internet connection (and a USB port) to start up the podcatcher client and receive the podcasts to which they have subscribed. Podcatcher on a stick <www.podblogger.de/mp3stick> "is a little program that can be installed/copied directly on a USB-MP3 Player. It has all the basic functions to receive podcasts like the other podcatcher-software, but instead of installing the software to a personal computer, podcatcher on a stick is installed directly on the audio-device (e.g. a MP3-USB-stick)."

The advantage of using 'podcatcher on a stick' is that the Podcast audio content is directly downloaded onto an audio device. Users do not need to worry about copying files manually or having music-management-software (like iTunes) installed. They merely connect the USB-MP3-Player to a computer, start the software, and let it download the podcasts directly to the player, and then disconnect the handheld device and start listening.

A newer program on the market is Audio Bay for Pocket PC <http://www.voiceatom.com/audiobay/>, an integrated podcasting solution for Windows Mobile Pocket PCs. Using the Audio Bay software, users can set up a podcast, record an episode, and upload it to the Web within a matter of minutes. While the quality of such a program may be limited, the immediacy possible could be a great asset for teachers who want to make class activities available as soon as possible.

An RSS Validator checks RSS feeds against the rules defined in the RSS 2.0 specification and validates elements of commonly used namespaces. To use an online validator, simply enter the address of the feed and click "Validate." If the validator finds any problems in the feed, it will provide messages for each type of problem and highlight where the problem first occurs in the feed. Using a validator can be necessary since, despite its relatively simple nature, RSS is poorly implemented by many tools. Using a validator is an attempt to translate the feed into code in order to make it easier to know when the RSS feed is being produced correctly, and to help users fix the code when it detects problems.

Popular online RSS feeds include RSS Scripting <http://rss.scripting.com> and FEED Validator for Atom and RSS <www.feedvalidator.org/>.

A new software package that expands the use of an RSS feed is Odiogo <www.pocketgear.com/software_detail.asp?id=19422>, short for Audio News to Go. This software converts the content on any Web site containing RSS feeds into small audio files using text-to-speech technology. Odiogo uses more natural speech patterns than previous versions in order to provide a better quality listening experience. Using Odiogo is like having a podcast of all favorite blogs. Users need to simply select their favorite feeds, click start, and then drag and drop the audio files into their handheld computer file and synchronize. Odiogo is designed to run on Windows XP/2000, but the MP3 files produced by Odiogo can be listened to on any PDA that plays MP3 files.

To create a podcast, one first needs to create an MP3 (or similar) audio file

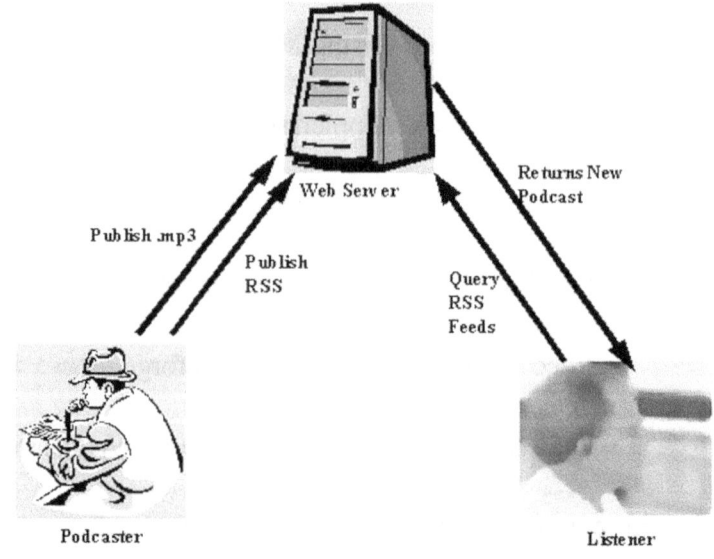

Figure 9:4 Podcast Signal

using audio editing software to edit files and include sound effects. Next, a podcaster will need to create the RSS file that accompanies the audio file, enabling users to subscribe to the recording and automate the download. In order to produce a podcast, authoring tools and Web service are necessary.

Some of the more common audio editors for Windows Operating System include:

- Adobe Audition® 2.0 <www.adobe.com/products/audition/main.html> approximately $250

- QuickTime® Pro <www.apple.com/quicktime/pro/win.html> $29

- Audacity <http://audacity.sourceforge.net/> Free

Audio editors for the Macintosh Operating System include:

- Final Cut Pro® <www.apple.com/finalcutstudio/finalcutpro/> approximately $1,299

- GarageBand™ <www.apple.com/ilife/garageband/> $79

- Audacity <http://audacity.sourceforge.net/> Free

An audio editor/player is needed to organize introductions, audio clips, or any bumper music the producer may have. Windows users may want to use an MP3 player with a lower memory requirement. WinAmp <http://winamp.com/> works well in this capacity and does not affect the audio recorder software. Macintosh users will want to stick to iTunes.

Macintosh is configured differently than a PC, and it is not as easy to capture audio at the same time a podcast is being produced. Audio Hijack Pro <www.rogueamoeba.com/audiohijackpro/> does a good job recording and enhancing audio. Those who decide to use Audacity to record podcasts will need SoundFlower and SoundFlowerBed <http://cycling74.com/products/soundflower.html>. SoundFlower is a Mac OS X (10.2 and later) system extension that allows applications to pass audio to other applications.

Once the audio has been edited, specific podcast software is necessary to publish the podcast for distribution. Apple's iTunes AAC format <www.apple.com/itunes/music/> allows podcasters to create "enhanced podcasts" complete with embedded photos at publisher defined points throughout the podcast. These files are only compatible with iTunes and iPods, leaving many listeners unable to take advantage of this feature.

To create a Windows Media enhanced podcast, the producer needs an application that supports Windows Media script editing. The free choice is Windows Media File Editor, which is bundled in the Windows Media Encoder download <http://www.microsoft.com/windows/windowsmedia/forpros/encoder/default.mspx>.

A complete list of software for publishing podcasts can be found at <www.podcastingnews.com/topics/Podcasting_Software.html>.

Once a podcast is complete, users will need a file transfer protocol (FTP) utility to upload the podcast to the Web host. Many people use Internet Explorer to download files from FTP sites. Many go to <http://tucows.com/> for FTP downloads.

The final step is to select a hosting site. Some may choose to host their own podcast on their school's Web server, while others may choose to use commercial podcast hosting sites such as BlogMatrix Podcast Hosting <www.blogmatrix.com/home/main/>. Their hosting plans start at less than five dollars a month and provide a complete blogging package for podcasting with unlimited monthly bandwidth.

Locating and Listening to Podcasts

With the thousands of podcasts available accompanied with a person's limited time, it can be mind boggling to find the exact podcast to meet one's educational needs. To access podcasts, users need either a Podcast search engine or a Podcast directory similar to the search engines or directories necessary for Internet searches.

PodSpider <www.podspider.com/website/en/podspider.php?popup=1> is one of the most popular Podcast Search Engines. In addition to a directory of podcasts, PodSpider offers client-side software to download and synchronize podcasts with MP3 players, not just iPods. Yahoo <http://podcasts.yahoo.com/> has also set up a vehicle to search for podcasts along with Podcast.net <www.podcast.net/> and IAmplify <www.iamplify.com/>.

Podcast Alley <http://podcastalley.com/> features the best Podcast Directory and the Top 10 podcasts as selected by the listeners. On this site, users will also find podcast software, the podcast forum, and a wide variety of podcasting information. In the same way, Podcast Bunker <www.podcastbunker.com/> provides a directory and tools for the average podcaster.

Audible.com, the popular source for online audiobooks is now providing Audible Podcasting <www.audible.com/adbl/site/preferences/podCasts.jsp?BV_UseBVCookie=Yes>. Audible listeners can receive all of their audio magazines, newspapers, and radio programs as podcasts in two different ways. Both methods require that users already have the appropriate software installed. The first method is to use a one-click subscription feature to add a personal podcast channel (RSS feed) to the podcast client. The second method is to "Copy and paste" a URL to add a personal podcast channel (RSS feed) to a podcast client.

The iPodder.org (www.ipodder.org) is a small program that runs on a desktop or laptop computer. It is organized into directories and its only purpose is to download audio files, usually MP3s, directly to an MP3 device. Currently iPod is supported on both Windows and Macintosh operating systems.

Free Government Information (FGI) is a location for initiating dialogue and building consensus among the various players (libraries, government agencies, non-profit organizations, researchers, and journalists) who have a stake in the preservation of free access to government information. Since an increasing number of governments and elected officials are finding podcasting a useful channel to inform their citizens of their views, FGI has developed a directory of government podcasts at <http://freegovinfo.info/node/174%20=>.

The Public Domain Podcast <http://publicdomainpodcast.blogspot.com/> consists of works of public domain authors read out loud in a weekly podcast. This Web site also includes links to public domain resources and topics of interest to literary and audiobook fans.

Many podcasters have set up podcast-specific RSS feeds identifiable with white-on-orange icons. To do this when a desired podcast is found, locate the links to the RSS feed necessary to add shows manually to a podcatcher application. To add the Web site's RSS feed, all one has to do is copy that URL and paste that information into the podcast-supported client's Add Feed window.

While it is possible to listen to a podcast on a desktop computer, a great deal more flexibility is available to listen to a podcast on a handheld device. To synchronize a Pocket PC with a podcatcher, the following steps need to be followed. There may be minor variations between podcatching programs.

1. Connect the handheld device to the desktop computers.

2. Start ActiveSync and select Tools->Options->Sync Options and in the list check the checkbox before 'Files'.

3. The Pocket PC has created a new subfolder in the 'My Documents' folder called 'Pocket_PC My Documents' or something similar.

4. Create a subdirectory in this new folder that will be used as a download-folder.

5. Put the path of this newly created subfolder in the first line of feeds.txt.

Manual synchronization works when the user desires only one podcast subscription, but it quickly becomes unmanageable with multiple subscriptions to synchronize on a regular basis. If a user subscribes to 10 or 15 podcasts, they can quickly add several hundred megabytes of new audio content to their portable player, sometimes on a daily basis. Therefore, automatic downloads is a better way to manage the process.

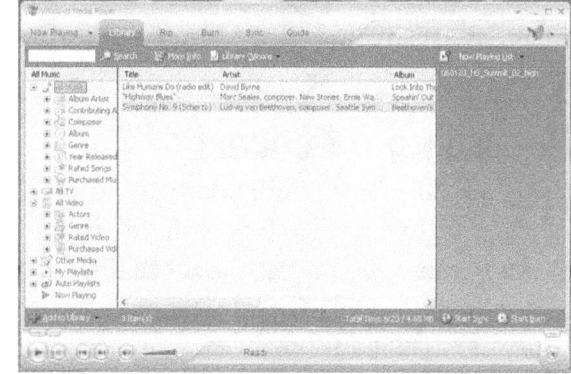

Figure 9:5 Screenshot: Windows Media Player Library

Once podcatching software is downloaded, the user needs to choose the default media players. If Windows Media Player is selected, click on the "Library" tab and expand My Playlists to expose this view.

If the user right-clicks on the playlist and chooses "Add to Sync List" from the menu, all the tracks are ready to synchronize with a portable device. With a portable player connected to the PC, change to the "Sync Tab" in Windows Media Player, click on the "Start Sync" button and the selected podcast files will download onto the portable device.

The most effective way to add podcasts to a portable device is an automatic download using Windows Media Player Auto Playlist to manage podcast subscriptions. While in the Media Player Library, right-click on the "Auto Playlists" header and choose "New" from the menu. After naming the playlist, add criteria by clicking on the plus sign under "Music" in My Library and selecting "More" from the drop-down list. Select "File Name" from the Choose a Filter dialog box, which defaults to Contains, and finally set the parameters.

In one swift maneuver, a playlist is created that contains everything in the *Podcast* folder and subfolders. Back on the Sync tab, with a little edit to the portable player synchronization options, the user is ready to download the latest podcasts every time he connects the player. With the handheld device connected to the desktop computer, display "Properties and Settings" and click on the "Settings" button on the Synchronize tab, making sure the player is set for auto synchronization by checking the box.

Eventually, synchronizing every podcast that is downloaded can become unwieldy. It is complicated to figure out which files will be listened to and which track is the latest podcast for a given subscription. By tweaking the Podcasts Auto Playlist, it is easier to keep track of new shows and avoid overloading the playlist. Parameters such as "Date Added to Library Is After Yesterday" to the playlist, limits synchronization to podcasts downloaded in the last 24 hours. This has the disadvantage of loading the playlist with every file from any podcast that was subscribed to since the day before. However, using this method the user will not have the entire back catalog of podcasts taking up most of the space on the media player. Each time he synchronizes the player, the playlist is refreshed with the latest tracks.

AOL has increased their podcast search functions in the WinAmp media player <www.winamp.com/player/>. WinAmp now has a podcast directory called SHOUTcast Wire <www.winamp.com/music/wire.php>.

For Macintosh and iPod users, Apple has created a podcast directory of more than 3,000 podcasts that can be accessed from their iTunes Music Store <www.apple.com/itunes/podcasts/>, allowing users to search by browsing through categorized lists. Podcasters can submit their shows via a link on the directory. Apple only links to the podcast files; it does not host them. To receive a podcast, users need simply to click the "Subscribe" button on the show's directory page, and it will automatically download to the user's iTunes library. Subscriptions are listed in a new podcast directory in the user's iTunes library, and new content is marked with a blue circle. PlaylistMag.com <www.playlistmag.com> is the ultimate iPod guide online.

Recording a Podcast

Every producer has a different way to organize a podcast. In developing a podcast, it is critical to remain true to the objectives of the show. By defining a clear purpose, the rest of the planning becomes easier because the podcast now has direction and focus. One of the most important parts of defining a purpose is to identify the audience.

Once the objectives are defined, students will need to decide what their main points will be, who will participate, what kind of script or outline they will need, and what kind of music will be appropriate.

After the planning segment comes the gathering of information and writing the script, identifying guests, locating Podsafe music, and finding a quiet location for recording. Next comes the practicing before the microphone, the actual recording of the podcast, and assembling its various components. At this point, a critical step that is often overlooked in the excitement of production is the final edit. The audio quality needs to be checked, any mistakes, noises, and filler words corrected, and any weak areas need to be re-recorded. The final step is to publish the MP3 file, generate a XML/RSS feed, check the MP3 link in the RSS feed, and publish the XML/RSS feed.

Audacity <http://audacity.sourceforge.net/> is a free, cross-platformed audio recorder and editor that is popular in the podcasting community. With Audacity students will be able to record live audio using their computer's sound input and then do simple editing to the recording and export it as an MP3 file.

After installing Audacity, an MP3 encoder is needed because of copyright and patent restrictions. The LAME <www-users.york.ac.uk/~raa110/audacity/lame.html> plug-in provides this functionality with just a few extra steps. In order for Audacity to use the LAME encoder, it needs to know what the path to it is.

Recording in Audacity, or other audio editors such as GarageBand, is similar to recording on a tape recorder but with more controls and visual information. When recording a podcast, one may find it necessary to record the show in segments. Every time a new recording is started, another track is created. Several audio sources can be used in a podcast. A recording of a voice or voices, sound effects, and music can be imported into Audacity. The sound levels of accompanying tracks can be changed over the course of the podcast and special effects can be added.

Podcasts are generally recorded and posted in either MP3 or AAC format for Macintosh computers. QuickTime 7 Pro on Mac OS X is an excellent program to record a podcast. One needs to make sure the audio input device is connected, and then go to "File" then "New Audio Recording." Next, click the red "Capture" button, begin recording, and click the black "Stop" button when finished. Finally, select "Export" from the File menu. In the Export dropdown box, choose "Movie to MPEG-4." Name the file with the .m4a extension (yourfile.m4a) in the "Save exported file as . . ." dialog box (this is required for compatibility with iPod). Select the desired location of the new file and click "Save." This method produces a file that is ready to be published.

> **If an M4A file is seen, users will know that it contains only MPEG 4 Audio. MP4 can be used for MPEG 4 video files, combined video and audio files, or just plain MPEG 4 audio. Both the .m4a and .mp4 container file formats are the same, they just have different file extensions.**

To record a podcast with GarageBand the user must create a Real Instrument track in GarageBand and adjust the Gain control on the audio interface to set the recording level of the speaker. Background music can be added in AIFF, MP3, or AAC file (except protected AAC files) format or a selection from GarageBand Apple Loops can be added. GarageBand Apple Loops allow the most flexibility since the length can be easily varied. Apple Loops helps users create their own unique copyright-free jingles for use in the podcast.

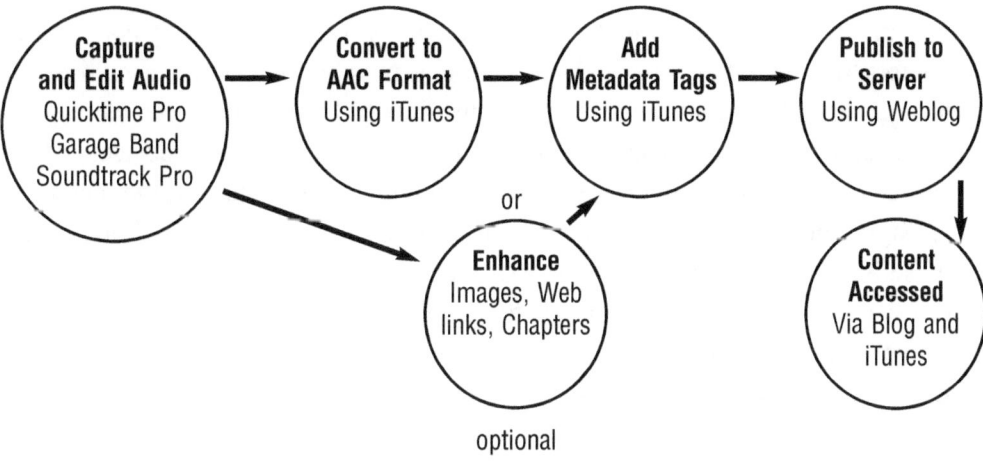

Figure 9:6 Workflow in Recording a Podcast with Garage Band

In recording educational podcasts, it is most helpful when students:

1. Follow a consistent format: day and date, welcome, introduction music, full introduction telling who the producer is, where they are located, why podcast, Web site location, location of Web site where podcast is located along with archive of old shows, announcements, news, closing remarks, and music.

2. Do not exceed 10 minutes in length.

3. An informal and high-energy tone is used.

4. The context for every post is provided along with a review of information presented.

5. Options for multiple learning styles are provided.

6. Flexible delivery options are provided so the podcasts can be played on a variety of handheld devices or desktop computers.

In recording a podcast, proper microphone techniques are vital along with a good microphone. Proper microphone techniques are a blend of physical positioning, diction, speed, mouth form, and presence, among others. The easiest problem to fix is microphone positioning. The speaker should be about a hand's width away from the microphone. In addition to the distance from the microphone, speakers can adjust their position relative to the microphone. Ideally, the microphone should be slightly above the speaker and to the left or right by up to 45 degrees.

When recording a podcast, care needs to be taken to reduce the noise. The brain filters out a lot of the noise to allow a person to concentrate on the more important sounds. However, microphones have no such filter, and podcasters may find that their recordings have picked up annoying noises in the environment. Two types of noise need to be addressed.

Environmental noise is the most common form of noise in recordings. This noise usually comes from fans, air conditioners, refrigerators, fluorescent lighting, and street noise. If possible, a podcaster should power off as many electronic devices as possible. If that cannot be done, they might try recording in a closet with lots of clothes to muffle the sound, or to pull a blanket over themselves and their recording device.

The other type of noise is *signal noise*, which is noise between the microphone and the recording device. With simple setups, signal noise can be caused by the use of an unshielded microphone cable. This is the main reason to use XLR cables, which have an extra lead to avoid noise from interference. Remember to use the shortest possible microphone cables to cut down on signal noise.

Talking too fast is a very common mistake for beginning podcasters. Therefore, it is advisable to practice the podcast several times before recording it for release. Another element of the speed problem is being tense. Audio recording, with the presence of a microphone and hearing one's own voice through the headphones, is alien to most students and it is difficult to be relaxed while doing it. Another cause for stress is unrealistic expectations. Neither students nor teachers should feel that they could sound professional at the beginning. The successful podcasters are generally happy as long as each show sounded and felt better to them then the previous show.

Hosting and Publishing a Podcast

After the podcast is created in MP3 format, an RSS feed needs to be created with an RSS generator such as listed in the previous section. The RSS generator gathers the necessary information for the RSS feeds and channel item elements. Options can be set along with the preferences that contain information about ownership, publishing settings, and audio compression. If the publishing preferences are set correctly, the software can publish the podcast and RSS feed by using the publish podcasts button.

RSS feed aggregators are extremely finicky and sometimes the feed is not coded correctly. A simple missed bracket or a misspelling will cause the feed not to work properly and will not be able to be downloaded. To solve this problem, after creating a feed, upload it to the server and then use an online validating tool such as Feed Validator <www.feedvalidator.org> to identify any problems. The feed validator will return a line-by-line analysis of any problems in the feed.

Adding an ID3 Tag to an audio file will ensure that the program is played. An ID3 tag is a file that is attached to an audio file containing album, artist, track, and other Machine-readable information.

Popular online sites to host podcasts include Poderator <www.poderator.com/>, Podomatic <www.podomatic.com/>, Podcast Alley <www.podcastalley.com>, Audioblogger <www.audioblogger.com>, Hipcast <www.audioblog.com/>, and FeedBurner® <www.feedburner.com>. A complete list of software for publishing podcasts can be found at <www.podcastingnews.com/topics/Podcasting_Software.html>.

The Apple Podcasting Server included in Mac OS X Server v10.4 makes it simple to publish and syndicate online content. From a server's point of view, a podcast can be simply a file attached to a blog. The server makes the blog and podcast freely available for subscription using the RSS 2.0 protocol. Individuals using Mac, Windows, and Linux can publish and access blogs using only their Web browsers without additional tools. Podcasts can be automatically opened in Apple's iTunes software on a Macintosh or Windows Machine.

If podcasters have access to their own Web server (or a shared server such as in a school), they can easily upload .MP3 content and an XML file that generates an RSS feed for podcasting clients. Using their own server, after they record their audio file, there will be two files to upload to the server, the audio file and the XML file telling the listener's computer when there is new content and the nature of that new content.

For those who desire to become seriously involved in podcasting, online communities are being set up to promote audio and video podcasts. One of the first online was PodcastPickle.com <www.podcastpickle.com/>, which provides an easy-to-use site to share podcasts and ideas to produce those audio and video podcasts.

Apple has set up an online community at iTunes Links <www.ituneslinks.com/> for bands, music fans, podcasters, and podcast listeners. The goal of this community is to help bands and podcasters get their music and message out to a much larger audience. Amateur school musicians can now have an outlet to share their talents without the expense of producing a CD that would have limited distribution.

Summary and Challenge

As soon as RSS feeds (Really Simple Syndication) technology developed and Web blogs began to gain popularity, innovators began to apply that same technology to audio files. Within months, audio blogs or podcasts were possible. At the same time, the iPod grew in popularity and users were not only able to listen to music, but also to audio broadcasting. Universities immediately started using this technology to disseminate class lectures and the K-12 environment began to follow their lead to find ways to incorporate podcasts into the curriculum in order to help raise test scores.

Teachers around the world are now finding podcasting an exciting tool to share content with their students and to have students share their work alongside professional broadcasters. Students can learn oral presentation skills and the technology necessary to produce a broadcast of their own. The resources of a classroom can now reach beyond the original classroom and become a resource for hundreds of other classrooms around the world.

Before going further, ask yourself these questions: "Would audio programming assist your students in raising their test scores? If so, where might you locate such programming?" "Will you be able to create you own podcasts?" "Will students gain from making their own podcasts? If so, what hardware or software will you need to produce your own podcast?"

chapter ten

Locating and Downloading eVideos and Vodcasts

While the technology to play video on a handheld device has been available for several years, it was considered a novelty and few handheld users took advantage of this feature. Within weeks of the release of the video iPod, news broadcasters, the television and film industries, educators, and amateur podcasters began releasing vodcasts for a wide variety of audiences.

The technology is so new that the terminology has not been standardized. The common terms are either "vidcast" for video cast or "vodcast," a term used for the online delivery of VOD (Video-on-Demand) using (Web/broad/narrow) casting content via RSS (Really Simple Syndication) enclosures. Other videos can be downloaded directly onto a handheld device without an RSS enclosure. These videos are file-based media that can be downloaded when the handheld device is synchronized.

History of Vodcasts

During its short history, video podcasts, or vodcasts, have made unbelievable growth.

- November 14, 2004, Steve Garfield created the first video blog, calling it a Video Podcast <http://stevegarfield.blogs.com/videopodcast/2004/11/videopodcast_20.html>.

- February 1, 2005, Gabe McIntyre taught the first course in a college curriculum on creating video blogging at the College of Arts in Utrecht, Netherlands.

- July 3, 2005, Dutch streaming pioneer Stef van der Ziel from Jet Stream introduced the term "VODcast" for channeling video-on-demand titles through RSS, XML, WML and HTML feeds at <www.vodcast.nl/index.html>.

- July 2005, The University of Chicago begins to vodcast poetry lectures and interviews with researchers.

- September 2005, international luxury lifestyle publication, *CITY Magazine*, becomes the first magazine to publish their *CITY TV* content on the iTunes Music Store as a vodcast.

- October 12, 2005, Apple introduces the new iPod and Front Row media center <http://www.apple.com/pr/library/2005/oct/12imac.html> iMac® with full support for vodcasting.

- Today there are hundreds of music videos, television shows, and Disney and Pixar short films at the iTunes Music Store for $1.99 each.

Sources for Educational eVideos and Vodcasts

Discovery Education provides unitedstreaming™ <www.unitedstreaming.com> video-on-demand subscription service containing more than 1,000 copyright-cleared video clips to teachers and students for editing or reproduction in class projects. Students may use the content of unitedstreaming in a bona fide educational or research project. Teachers and librarians can help students select a relevant video and download the video clip onto their handheld device.

DigitalCurriculum™ <www.digitalcurriculum.com/> from Discovery Education is also a curriculum video-on-demand teaching and learning system that provides full-length educational videos and key concept video clips, audio, graphics, text, and images to subscribing teachers and students. According to their Terms of Use, these digital formats may be transferred to CD, DVD, or other physical media for student and teacher multimedia projects or student and teacher educational (and non-commercial) presentations only. This agreement broadens educational usefulness as it allows students or library media specialists to download videos onto handheld devices for viewing at their convenience.

PowerMediaPlus.com <http://powermediaplus.com/> provides students solid connections to core concepts with media-on-demand. This digital media program of 2,600 videos allows educators to easily integrate standards-based multimedia into their school and curriculum. These video clips are precleared for user-directed editing. Using a school's own software, librarians, teachers, and students can shorten or lengthen clips, renarrate the programs, or incorporate the clips into their own multimedia projects.

LearnOutLoud.com <www.learnoutloud.com/> provides more than 500 free audio and video titles. This directory features free audiobooks, lectures, speeches, sermons, interviews, and many other free audio and video resources. Most audio titles can be downloaded in digital formats such as MP3 and most video titles are available to stream online. Streaming media generally have copyright protection and cannot be archived. Copyright cleared audio and video can be stored on desktop or handheld devices to be used by a large number of students.

Google's video store <http://video.google.com> is less controlled than Apple's iTunes Music Store <www.apple.com/itunes/videos/>. Google prefers the phrase "the first open video marketplace." Google video allows anyone to post from the biggest TV network to the most talent-free camera-phone owner. Since there is little control of the types of media made available at this site and some unacceptable videos may be posted, parents and educators need to monitor the video downloads from Google video.

Google has digitized more than 100 historic video clips for the National Archives <http://video.google.com/nara.html>. This is a free online service available to teachers, students, and the public. Motion picture films, primarily from the 1930s, document the history and establishment of a nationwide system of national and state parks. In addition, this collection also includes 1970 film documents of the expansion of recreational programs for inner-city youth across the nation.

Other historical videos, documents, and online exhibits are available at the U.S. National Archives and Records Administration <www.archives.gov/>. Google said it is exploring the possibility of expanding the project to include more video on its Web site.

The videos from iTunes Music Store always cost two dollars an episode and every show is downloadable and transferable to an iPod. The video quality of color and clarity is good, often with professional production values, and there are no advertisements.

At Goggle's video store, some videos are copy-protected; others are not. Some can be downloaded, while others can only be viewed online. There is a wide variation of resolution and production quality at Goggle and some videos include commercial advertisements. Some videos are free and some cost money.

For free and legal MPEG-4/DIVX video content for a handheld device, Public Domain Torrents <www.publicdomaintorrents.com/> provides an interesting selection of resources. This site contains older and obscure movies that are in the public domain. All the files are in bit torrent format and most are offered in various screen sizes and frame rates. Works become part of the public domain, meaning no longer owned by their creators, when they reach a certain age and/or when the original creator/owner does not renew the copyright. All the movies listed on this site are believed to be in the public domain. If anyone believes something listed on their site is not public domain, a link is provided to email the site owners to render an objection.

One of the more popular exclusively audio/video search engine is Singingfish™ at <http://search.singingfish.com>. Singingfish only indexes multimedia formats, including Windows Media, Real, QuickTime, and MP3s. Video search engines are available at Yahoo <http://video.search.yahoo.com/>.

Blinkx TV <http://tv.blinkx.com/> is a search engine that allows users to search for audio and video clips using not only keywords and phrases, but also the content in the actual clips. The Date/Relevance Slider lets users better match their interests with the content available. Blinkx.TV maintains RSS Support, which means users can save any search as an RSS feed so they will be automatically alerted every time content relevant to their search appears on the video Web.

PocketMovies.net <www.pocketmovies.net/> offers MPEG-1 encoded trailers for playback using PocketTV <www.pockettv.com/>. They also offer a search engine for movies that are compatible with PocketTV.

TiVoToGo Mobile Service is now available for Windows-based handheld devices. Until now, owners of the TiVo Series2 digital video recorder could transfer their TiVo shows to Windows-based computers including laptops. However, with the latest update, users can watch recordings on Windows Mobile-based Portable Media Centers, smartphones, and Pocket PCs.

Selecting Software and Media Players for Video Playback

In recent months, consumers have been turning from single function devices to all-in-one devices containing PDA/video/audio and sometimes even telephone features. A media handheld device means that the device has at least one SecureDigital slot and users can use different SD cards at different times, each with its own style of music, audiobook, or video stored on them.

The 4-inch PocketDISH™ <www.pocketdish.com/> and the 2.5-inch video iPod both are equipped with a 30 GB hard drive. However, the PocketDISH is a more powerful media recorder that eliminates the need for multiple devices and can store up to 120 hours of video. One of the biggest advantages of the PocketDISH over the iPod is the cost of the TV shows. With PocketDISH, the cost to transfer content is free, and users do not have to wait a day to download the broadcast.

Figure 10:1 PocketDISH

The new Apple Video iPod will support up to 150 hours of video on a 2.5-inch color display with up to 20 hours of battery life. Videos for the iPod easily can be downloaded directly from the iTunes Store for $2 a program. Users can subscribe to and view some video podcasts with iPod-specific feeds. However, users cannot grab just any video podcast and drop it on an iPod. iTunes' inability to convert existing video into an iPod-friendly format is its biggest weakness. To make existing video into something that can be played on the iPod, users have to convert the video using either Apple's own $30 QuickTime Pro application or another video encoder that can work with the H.264 and MPEG-4 formats that iPod uses.

The new iPod with video support is great, but unfortunately, it can only play the videos bought from iTunes, and home movie files were not in the right format until iTunes 6.0.2 was released. The iPod supports MOV, MP4, and M4V file formats at up to 768 Kbps, 320-by-240 pixels, and 30 fps.

To transfer a home movie onto a Video iPod the movie must first be in .mov format before it can be dragged into an iTunes video library. After the video is imported, users should see a small thumbnail of it in their video library. At this point, users merely need to right-click on the video they would like to have on their iPod, select "Convert Selection for iPod," and wait for iTunes to convert the file. This conversion can take anywhere from just a few moments to several minutes, depending on the file size. When iTunes has finished, users will see a new thumbnail of their video in the iTunes video library on their desktop. The next step is to synchronize the iPod to iTunes and the video should be on the iPod ready to enjoy.

Figure 10:2 Video iPod

While Microsoft's Windows Media Player 10 can automatically convert videos into a format suitable for watching on a portable media player, third-party applications may be necessary to convert videos to a playable format for the iPod. However, such applications are not prevalent, and they are a cumbersome extra step to getting video into a portable media player.

Windows Media Player 10 Mobile is available on Windows Mobile-based devices running Windows Mobile 2003 Second Edition or later. Microsoft Media Player 10 supports the most popular file formats including Windows Media Audio (WMA), MP3, and Windows Media Video (WMV). It also adds support for Microsoft DirectShow® technologies along with support for other developing media formats.

To make sure the digital music and video desired will play back on a specific device every time, look for the PlaysForSure logo. Match the PlaysForSure logo <www.microsoft.com/windows/windowsmedia/playsforsure/> on a large selection of leading devices and online music stores.

Flash® Professional 8 authoring software <www.adobe.com/products/flash/flashpro/> features new tools for embedding video converted from popular digital video formats and compression systems. New plug-in tools allow digital video producers to encode to the Flash FLV (Flash Video) format directly from most video editing programs. Flash's video encoder is also available as a stand-alone program for batch video encoding. Flash 8 can compress QuickTime and AVI files to Flash's proprietary FLV format, without a major loss of audio or display quality. Flash 8 also includes a new and impressive set of tools for authoring graphics and animation, including a preview-and-testing environment for mobile devices, and support for the MIDI audio format that many mobile devices use.

Flash Lite™ and Flash Player SDK can be incorporated into mobile devices by manufacturers, allowing users of those devices to view Flash animations in the same way as desktop PC users. In early 2005, Adobe acquired Macromedia, the developers of Flash. Those developers claim that creating a mobile application using Flash can be three to five times faster than using alternative technologies. Adobe reported that shipments of mobile devices that included Flash increased from 12 million units to 45 million units during 2005 and devices featuring Flash Lite will soon follow.

The RealPlayer for Mobile Devices features playback of RealAudio, RealVideo®, and 3GPP compliant content via streaming or download. It provides seamless integration with the RealPlayer for the Pocket PC to drag and drop MP3 and RealAudio files to devices. Access to news, sports, movies, music, and radio content is available via RealPlayer Mobile Media Guide.

PictPocket Cinema by DigiSoft <www.digisoftdirect.com/products/pictpocketcinema.html> "is a unique multi-media viewing program that has added a desktop converter that allows users to display AVI or QuickTime movies with unsupported video codecs on a Pocket PC."

Nevo® <http://www.mynevo.com/html.php?page_id=21>, by Universal Electronics Inc., is a suite of technology that enables complete audio visual, home, and digital media control. It includes the Nevo Client software for handheld devices, NevoMedia Server, and NevoMedia Player that allows users to connect seamlessly, control, and interact with digital media and consumer electronic devices in the networked home. Users are able to transfer media from one device to another seamlessly without having to use another control. NevoMedia Player embedded on an iPaq handheld computer can support the following file types: Audio - .wma, .MP3; Video - .wmv, .wmv (copy protected); and Pictures: .bmp, .jpeg, .gif, .png.

PocketStreamer Pro from Handago.com <www.handango.com> provides LIVE TV, radio, music, news, and weather for the MP3, wma, or wmv formats. PocketStreamer Pro is compatible with Pocket PC 2003, 2002, 2000 and Windows Mobile 5. To take advantage of the PocketStreamer the Pocket PC needs to have Internet access and Windows Media Player.

The Core Pocket Media Player (TCPMP) <tcpmp.corecodec.org/download> is a free audio/video player that supports a variety of audio and video codecs not supported by Media Player Mobile, including DivX, XVid, or MPEG video, and Ogg Vorbis and MP4 audio formats. The program is available for both the Windows Mobile and Palm OS devices.

Converting Video Formats to Play on a Handheld Device

Changing video formats so it can be played on a particular handheld device can become extremely confusing. While many educators would prefer to leave such details to the computer technicians and "geeks," they often find themselves losing considerable time trying to figure out the correct software and setting to use. Memorizing data that will be soon forgotten is generally not productive, but understanding the basic principles involved and a few well-selected bookmarks can become invaluable.

Fortunately, there are software packages that help with the conversion of video formats from full screen to miniature screens. In order to put home movies, DVDs, or TV programs on a handheld device and watch them in great quality, with stereo sound and in full-screen, landscape mode, a memory card as small as 128 MB is sufficient to store a full-length feature film, or TV shows and home movies up to 100 minutes.

With Pocket DVD Studio 3.5 <www.pocketgear.com/software_detail.asp?id=14059> users are able to watch DVD movies or TV shows on Pocket PCs, Palm, smartphones, or portable media center operating systems. This conversion program claims to be 300 percent faster in high quality and supports video formats or TiVo, WMV, DivX, MPEG, RM and many more for the Windows Mobile operating system. Unlike other software, Pocket DVD Studio works in one single step. It reduces intermediate conversion steps, saves time, and produces better audio/video quality with no loss in picture details.

MoviesForMyPod (formerly MovieToGo) for the Mac OS X, Panther, Tiger, G3 and higher, is a simple batch processing application for converting QuickTime movies into iPod friendly 320 x 240 H.264 movie files. The Macintosh program, available as a free download from the Digigami Web site <www.digigami.com/download/>, makes it easy to convert a collection of QuickTime movies "To Go" with a video iPod. Converting movies to the Video iPod format can be done by simply opening the movie file in MoviesForMyPod, clicking on the "Convert to Video iPod" icon, naming the new movie file, and letting the software do the work. After the file conversion is complete, the user simply double-clicks on the new movie file and the video will automatically be added to the iTunes Video library.

Other video converters include:

- PQ DVD to iPod Video Converter <www.downloadatoz.com/dvd-to-ipod-video-converter/> is a one-click, all-in-one solution to convert DVD, Tivo, DivX, MPEG, WMV, AVI, RealMedia, and many more to iPod Video.

- PSP Movie Creator <www.pqdvd.com/> helps users to create/convert PSP® video/movies in one click. The software converts DVD movies and popular video formats to Sony PSP.

- PQ DVD to iPod Video Converter <www.mp3towav.org/PQ-DVD-to-iPod-Video/> converts DVD to iPod Video in one click, as well as Tivo2Go, DivX and other popular video files to iPod.

- Videora iPod Converter <www.videora.com/en-us/Converter/iPod/> is a free video conversion application that allows users to convert their PC video files (avi, mpeg, etc.) into the proper video format that the iPod understands.

- Cucusoft iPod Video Converter <http://www.cucusoft.com/> is an easy-to-use video converter software for Apple iPod Movie and iPod Video.

- Pocket DVD Wizard 4.7 <www.pocketgear.com/software_detail.asp?id=13688> converts high quality video and stereo sound DVD to Pocket PC.

- Pocket DVD Studio 4.0 <www.pocketgear.com/software_detail.asp?id=14059> provides a one step method to convert DVD and video files to be played on handheld devices. This software reduces intermediate conversion steps that saves time and produces a better video/audio quality with no loss in picture details.

- DVD to Pocket PC 3.6.2 Software <www.makayama.com/dvdtopocketpc.html> converts the content of a video file or DVD to a super small movie file, which will play on any Windows Mobile 2003 device on a postage stamp size memory card.

Shooting Video for the Micro Screen

To shoot video for the handheld screen, the most important issues are to keep it close, less complicated, and movement to a minimum. Establishing shots should be tighter and less frequently used. The use of special features in the editing software can be distracting and difficult to decipher, and complicated backdrops should be avoided. Stick to simple cuts whenever possible.

Even though video compression has improved, movement can still be a problem resulting in stuttering pans and blocking transitions. Placing the camera on a tripod and only using slow pans and tilts when the story needs them will help remove the shakiness and simplify the compression process. When developing the script, remember that the movement of a subject within the frame could be affected by compression.

Preparing a Vodcast

Preparing a vodcast is a simple procedure using basic hardware and software available to the educational consumer. The technology behind a vodcast is the same as those of a podcast. The RSS feed is the common thread of an audio podcast, a video vodcast, and a text blog.

The first step in preparing a vodcast to be played on an iPod is to create the video. One helpful resource to aid in preparing video for podcasting is a book from Linworth Publishing entitled *Creating Digital Video in Your School* by Ann Bell.

Once the video is prepared, it must be digitally compressed to conserve storage space. A helpful tool to do this is QuickTime Player Pro 7 <www.apple.com/quicktime/pro/win.html>. To get a great looking video for the small screen size, the video needs to be saved in the H.264 format for an iPod or in MP4 format to be played on Pocket PCs. To keep file size small, select a modest image size such as 240 x 180 and a frame rate of 15 fps.

The next step is to upload the video to the server from which the movie will be delivered and place the compressed movie file in a download directory on the Web site.

If a person does not have access to a server, he can use iDisk <www.apple.com/dotmac/>. iDisk provides storage space on Apple's secure servers where users can invite colleagues to access their files, whether text, audio, or video. A list of other free online storage spaces can be found at Free Online File Storage <www.lights.com/pickalink/freestorage/>, but RSS feeds may not be able to be set up to a particular site. In those cases, access to the video will be limited to spreading the word among friends and contacts that a new video is available.

After the video is on a server, an XML file needs to be created that will allow iTunes to access the movie. The XML file needs to be placed in the same directory as the movie. If iDisk is used, place the XML file in the sites folder.

When all the components are assembled, it is time to test the video in the browser by launching the browser and entering the address to the XML file in this form—<www.yoursitename.com/yourvodcast.xml>. A page should appear that lists the title of the vodcast. The URL in the address field should have changed so that feed replaces http, as in: <feed://www.yoursitename.com/yourvodcast>.

Finally, producers will want to check the vodcast in iTunes to make sure everything is working correctly. To do this, launch the current version of iTunes and select "Subscribe to Podcast" from the "Advanced" menu. Enter the desired URL and then click "OK." An entry for the vodcast should appear in the Podcast playlist within the iTunes main window and the video will begin downloading to the computer. Once the producer is assured that the vodcast is exactly what they desire, it is time to announce it to the desired audience.

Another quick and easy tool for publishing audio and video podcasts is VODcaster <www.twocanoes.com/vodcaster/>. VODcaster allows users to enter in all the important information without having to know XML.

Any video editing software package can be used to create a video for a VODcast, but one software in particular was created strictly for VODcasting. Vlog It software from Serious Magic™ <www.seriousmagic.com/products/vlogit/> lets users create newscast-like blog entries via a simple, drag-and-drop interface. Users can insert music or video clips into a vertically scrolling timeline, and then use a camcorder or Webcam to record their video and a microphone to add voiceover. The Vlog It timeline has an integrated teleprompter for displaying the narration. If a user has a green background, she can use composites to add a video or animated backdrop of her choice with a chromakey effect similar to what weatherpersons do. Once the video blog is created, the software automatically adjusts the compression and codec settings based on the selected output choice. Vlog It can even upload the file to a video-hosting service.

One of the problems in preparing a podcast or a vodcast is the multitude of possible file formats that could be used. To help solve this problem, a good conversion tool can make a real difference. One possibility is MediaCoder <www.jakeludington.com/downloads/20060222_mediacoder.html>, which can unify many of the readily available open source conversion tools into one interface with many preconfigured output options to simplify audio and video conversion.

Summary and Challenge

Preparing educational videos and sharing them over the Internet has been popular for a number of years. Now, with the advent of handheld devices with the capacity to play video, the potential for students and teachers to share video is exploding at all grade and curriculum levels. Basic digital videos can be emailed or posted on a Web site, but with the expansion of blogs to include audio and video broadcasting, users can be notified by RSS feeds whenever a new program is available. Because video podcasts can be produced by all levels of equipment, from the extremely basic consumer level to the professional level, educators easily can find hardware and software that would best meet their instructional needs.

Evaluating the wide variety of eVideos and vodcasts is key for a student to gain understanding of media literacy and meeting the various aspects of multiple intelligences. Personal digital media players are quickly being transformed from centers of entertainment to centers of education.

Before continuing, ask yourself the following questions: "Will curriculum-on-demand work in my particular educational environment?" "Will it be possible to download eVideos or vodcasts on my students' handheld devices?" "How might students prepare eVideos that enhance their multiple intelligences and help them raise their test scores?" "What additional equipment and training will you need to circulate eVideos or help prepare student-produced vodcasts?"

chapter eleven

Digital Media Copyright Issues

The fundamental base of American copyright law has always been the attempt to find a balance between the necessity of protecting the rights of authors in order to encourage production of intellectual works, and the necessity of providing public access to works in order to maintain a democratic and educational society. With the expanding use of technology, it has become extremely complex and difficult to maintain adequate legislation.

Fair Use Guidelines

All educators need to be familiar with the Fair Use Law, Title 17, United States Code, Public Law 94-553, 90 Stat. 2541, which gives citizens special exceptions to the strict legal copyright requirements. Fair use provisions of the copyright law grant users conditional rights to use or reproduce certain copyrighted materials as long as the reproduction or use of those materials meets defined guidelines. In order to determine if a specific situation falls within the Fair Use Guidelines four factors need to be considered:

1. "The purpose and character of the use, including whether such use is of a commercial nature, or is for nonprofit educational purposes;

2. The nature of the copyrighted work;

3. The amount and substantiality of the portion used in relation to the copyrighted work as a whole; and

4. The effect of the use upon the potential market for or value of the copyrighted work (17 USC, §107)."

Berne Convention

The Berne Convention is an international agreement about copyright, which was first adopted in Berne, Switzerland in 1886 to establish the Protection of Literary and Artistic Works <www.copyright.gov/title17/92appii.html>. In 1989, the United States became a party to the Berne Convention. Signers of this treaty agreed to protect the copyrights of the others under a country's own laws. The agreement made copyright enforcement easier because a person only needs to know the copyright laws of his own country rather than those of hundreds of other nations. With the sharing of podcasts and vodcasts via the Internet, this treaty becomes important for media producers.

Visual Artists Rights Act

Title 17 USC Section 106A is known as the Visual Artists Rights Act of 1990 (VARA) <www.copyright.gov/title17/92chap1.html>. This act was the first time in federal law that an artist's moral rights in her works of art were recognized. This act covers the creation of paintings, drawings, prints, or sculptures, existing in a single copy or in a limited edition of 200 copies or fewer.

VARA grants artists two new rights, the right of attribution and the right of integrity. The right of attribution concerns the artist's right to claim authorship of a work created by him and to deny authorship of a work not his own. The right of integrity concerns the artist's right to prevent or to recover damages for the intentional distortion, mutilation, modification, or destruction of his work. The revolutionary aspect of VARA is that the artist retains these rights throughout his lifetime, even when the original work to be protected is no longer in his possession.

Fair Use Guidelines for Educational Multimedia

After the 1976 Copyright Act was enacted, a great explosion in the use of educational technology occurred that was not specifically addressed in that act. To help provide guidelines that are more specific so that educators could follow them and be reasonably sure that they would not be in violation of the copyright law, the *Fair Use Guidelines for Educational Multimedia* was developed by the Consortium of College and University Media Centers (CCUMC) <www.ccumc.org/copyright/ccguides.html>. These guidelines were made a part of the Congressional Record and became an unrelated part of a Judiciary Committee Report. These guidelines do not represent a legal document, nor are they legally binding. They do represent an agreed-upon interpretation of the fair use provisions of the Copyright Act by the overwhelming majority of institutions and organizations affected by educational multimedia.

These guidelines state <www.ccumc.org/copyright/ccguides.html#uses>: "Educators may incorporate portions of lawfully acquired copyrighted works when producing their own educational multimedia programs for their own teaching tools in support of curriculum-based instructional activities at educational institutions." For student use these guidelines state: "Students may incorporate portions of lawfully acquired copyrighted works when producing their own educational multimedia projects for a specific course."

Uses of educational multimedia projects created under these guidelines are subject to the time, portion, copying, and distribution limitations listed in Section 4.

Educational Media Fair Use Guidelines	
Time	Educators may use their multimedia projects for teaching for a period of up to two years.
Portion	Amount of copyrighted work that can reasonably be used. These limits apply cumulatively to each educator's or student's multimedia project for the same academic semester, cycle, or term.
Format	Portion Used
Motion Media	Up to 10% or 3 minutes, whichever is less.
Text	Up to 10% or 1,000 words, whichever is less.
Music, Lyrics, and Music Video	Up to 10%, but not more than 30 seconds.
Illustrations, Cartoons, and Photographs	May be used in its entirety, but no more than five images per artist or photographer.
Numerical Data Sets	Up to 10% or 2,500 fields or cell entries, whichever is less.

Table 11:1 Fair Use Guidelines for Educational Media

The Digital Millennium Copyright Act

In October 1998, *The Digital Millennium Copyright Act (H.R. 2281)* was signed into law. This law protects copyrighted works, both digital and print, against unauthorized copying using new forms of technology. The basics of the DMCA include:

- "The DMCA prohibits the circumvention of any 'technological protection measure' (such as a password or form of encryption) used by a copyright holder to restrict access to its materials.

- The DMCA prohibits the manufacture of any device, or the offering of any service, primarily designed to defeat a 'technological protection measure.'

- The DMCA exempts any online service provider (OSP) or carrier of digital information from copyright liability based solely on the content of a transmission made by a user."

Digital Performance Rights in Sound Recordings Act

In 1995, the *Digital Performance Rights in Sound Recordings Act* <www.copyright.gov/legislation/pl104-39.html> was passed to limit the digital transmission performance of a sound recording. Digital transmission includes Internet transmissions and certain digital satellite transmissions. Therefore, podcasters need to take special care in selecting music for their programs.

TEACH Act

In 2002, the *Technology, Education and Copyright Harmonization Act* (TEACH Act) <http://frwebgate.access.gpo.gov/cgi-bin/getdoc.cgi?dbname=107_cong_bills&docid=f:s487es.txt.pdf> established the rules under which copyright protected materials could be used in online education. The law anticipates that students will access each "session" within a prescribed period and will not necessarily be able to archive the materials or review them later in the academic term; faculty will be able to include copyrighted materials, but usually only in portions or under conditions that are similar to conventional teaching and lecture formats.

Software Downloading Regulations

Downloading from the Internet does not mean all the programs are free. Downloaded software can be divided into four categories.

- Public domain programs and files carry no copyright and have no limits on redistribution, modification, or sale.

- Freeware programs and files are free to use and give away, but not to sell or modify. The author retains the copyright.

- Shareware programs and files allow users to road-test programs for a short evaluation period, and then you must either pay the author a small fee or erase the program from the computer. The author retains all copyrights.

- Commercial software is software sold for a profit.

Summary and Challenge

With the fast changes in technology, it is difficult for Congress to keep pace with appropriate copyright laws that protect the intellectual content of the producer as well as provide fair use for those wishing to use media for educational purposes. As media is transferred instantly around the world, steps need to be taken so that both users and producers of media abide by the appropriate copyright laws, and they teach their students to do the same.

Copyright is a complex issue with many legal technical issues. In order to help educators sort through the legalities, a multitude of copyright resources are available, both online and in print format. One of the better ones to consult is the fourth edition of *Copyright for Schools: A Practical Guide* by Carol Simpson from Linworth Publishing.

Before continuing, ask yourself these questions: "How have copyright issues affected how you teach your students?" "Where can you find the best answers to your copyright questions?" "Where can you find public domain resources?" "Who do you contact in order to receive copyright clearance?"

chapter twelve

Incorporating eBooks, eAudio, and Podcasts into the Curriculum

Put the right tools and the right training in an educator's hand and she will find a way to incorporate them into the curriculum. The availability of handheld devices, the software, text, and media that can enrich their curriculum is an exciting field that is emerging for teachers and media specialists.

Collaboration

Beaming from one handheld computer to another is one of the most overlooked features of handhelds. However, teacher collaboration can be greatly improved using this feature. Software that simplifies the process of beaming between different operating systems includes Documents to Go <www.dataviz.com/> and Quickoffice® <www.quickoffice.com/>. These programs are compatible with the desktop versions and can easily be shared among users.

Teacher-Produced Assignments

Some teachers like to personalize assignments by loading sections of books and magazine articles onto students' handheld devices. They can add notes, organizers, comments, and questions before converting a text to an eBook format. Such personalized assignments can be extremely helpful for students with special learning needs or those who are unable to be in class every day.

Reading Strategies with eBooks

One of the first ways many educators begin to have their students use eBooks is to download a required reading book onto the student's handheld device, but that is only the beginning of its possibilities.

While the text of an eBook may be displayed on an electronic device, its use and application are not much different from paper-based materials. The use of eBooks applies comfortably in many reading strategies teachers use to assist students in improving their reading skills. Dr. Terry Cavanaugh, a visiting assistant professor at the University of North Florida's College of Education and Human Resources developed the eBRS Program <www.drscavanaugh.org/ebooks/ebrs/intro.htm> to assist students in the metacognitive process as they learn how to learn by providing active strategies to help them organize, understand, and learn material.

One advantage of using reading strategies with eBooks is that all the reading and learning activities are used within the eBook and no other materials are needed, saving costs on books, paper, and pencils. Varying the types of reading strategies helps students improve their standardized test scores to meet No Child Left Behind accountability goals.

Dr. Cavanaugh's eBook Reading Strategies on his Web site <www.drscavanaugh.org/ebooks/ebrs/intro.htm> include:

1. "activate background knowledge

2. set a purpose for reading

3. help in identifying the main idea

4. support the main idea with clear, complete explanations

5. organize information

6. increase comprehension of vocabulary and concepts

    7. develop their metacognition process"

Dr. Cavanaugh notes additional reading support strategies that are available using the eBook format. The terms in [brackets] indicate an eBook support format.

- "Take notes while reading to assist in understanding [Note/text box]

- When text becomes difficult, read aloud to assist in understanding [Text-to-Speech]

- Underline or mark-up information in the text to assist in remembering [Highlighting]

- Use reference materials (dictionaries, etc.) [Interactive dictionaries/Hyperlinks]

- Paraphrase/restate to better understand

- Go back and forth in the text to find relationships among ideas [Bookmark]

- Ask self-questions to find answers in the text"

While no one will dispute the importance of traditional print material, digital text and eBooks provide additional reading supports that are not available in print format. Some of these added features include:

- Copy and paste out notes

- Read aloud text

- Just-in-time word look-up with interactive dictionary

- Hypertext to Web sites for additional information

- Adjust size of text for readability

- Auto summarize text

- Print marked selections

- Auto index

- Search within a document for specific text (find-find next)

Cross-Curricular Uses of eBooks

One of the current trends in podcasting is "sound seeing," in which students or teachers record narrations of their travels. For example, before a ninth grade biology field trip to a game reserve, the class studied each of the species they might see. After gathering their information, each group of five prepared a podcast on a different animal. Before leaving the school for the game reserve, each student loaded all the podcasts onto their handheld devices. While touring the game reserve, the students listened to the podcast about the animal as they observed it in the game reserve.

eBooks in Language Arts

A seventh grade English teacher might select eBooks from Microsoft Reader Catalog of eBooks <www.mslit.com/> specific to the reading level and interest of each student. While reading the books, students can highlight and bookmark important ideas and information. The color of the highlighting can be changed for different topics or themes of information.

Using Microsoft Reader, students are able to add annotations to their eBook expressing their reflections on a particular passage. A single annotations file saves all the notes automatically, referencing the location of the notes within the book.

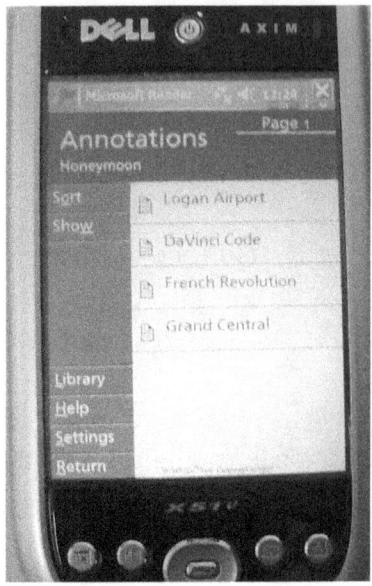

Figure 12:1 - Annotation Page of Microsoft Reader on Handheld Device

Every student's handheld needs to have a dictionary, such as the *Encarta® Pocket Dictionary* <www.mslit.com/details.asp?bookid=MSMSEBDICT>, loaded on it. When a student comes upon a word that she is unsure of within the text of a story, she can look up the word and attach a note to the unknown word for future reference.

For students with limited reading, a text-to-speech plug-in <www.microsoft.com/reader/developers/downloads/tts.asp> will read the text to the student as they read along. Currently the Microsoft Reader 2.0 is only available for Windows-based PCs and laptops or Microsoft Reader 2.0 for Tablet PCs.

In freshman English, students can write their own short stories on a desktop computer. After the stories have been edited, they can be converted to Microsoft Reader format using The Microsoft Reader (RMR) plug-in <www.microsoft.com/reader/developers/downloads/rmr.asp>. After this plug-in is downloaded, a new listing of READ is added to Word's FILE menu, which will temporarily convert a Word document into a Web page and then into an MS Reader eBook. Students can then synchronize their handheld devices to share each other's short story.

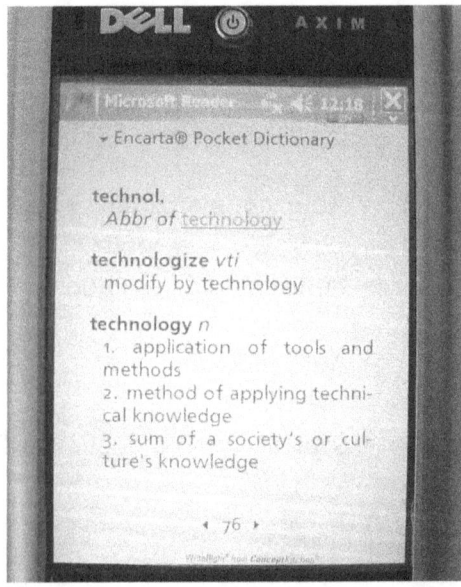

Figure 12:2 - Dictionary Page of Microsoft Reader on Handheld Device

eBooks in Foreign Languages

Other disciplines can also find the use of handheld devices supportive of their curriculum standards and benchmarks. For example, all students in a foreign language class will find it helpful to have a foreign language dictionary, such as *English-Spanish Pocket Dictionary* <www.mslit.com/details.asp?bookid=MSDICTENES>, loaded on their handheld. Whenever a new word is needed, that word can be bookmarked for easy reference. Individual tests could be prepared from the new words that were consulted in the dictionary.

eBooks in Science Curriculum

With the fast changing pace of developments in science, a freshman Biology Class might use eTextbooks from McGraw-Hill <http://mhln.com>. The students preload the necessary assigned reference pages on water quality to their handheld devices. The class assembles the water-quality probes that are compatible with their handheld devices before the class field trip to the nearby Mississippi River. While onsite, the students compare the water quality in the river with the data available in their textbook and record their observations in the word processor on the handheld computer.

Several scientific probes are available for school use. The ImagiProbe Wireless System <www.imagiworks.com/> provides a powerful software application and a sensor interface, which when combined with sensors and a handheld computer, creates a portable laboratory that enables students to collect and visually analyze data.

Vernier Technologies <www.vernier.com/> developed Data Pro, a data-collection program for use with Palm-powered handhelds. Users can load Data Pro software onto their handheld, connect the handheld to the LabPro (with a Handheld-to-LabPro cable and cradle), attach a Vernier sensor, and begin collecting data!

Pasco Probes <www.pasco.com/> developed the PowerLink interface that combines the ease-of-use of the PASPORT line with the convenience of a USB hub. When "C" size batteries are added, students can collect data remotely using a laptop or Palm handheld.

Concord Consortium's <www.concord.org/work/themes/handhelds.html> most recent work is focused on probeware for handheld computers. They have developed new probeware software called CCProbe, which is built upon the open source CC LabBook System. CCProbe works on Palms and Pocket PC handheld computers as well as Mac OS, Windows, and UNIX desktop operating systems.

eBooks in Social Studies Curriculum

Instead of subscribing to multiple copies of the local paper so that each student can discuss the current events, a 12th-grade civics teacher might subscribe to the online service AvantGo <www.avantgo.com>. The teacher selects the appropriate news channels to download, and each day the students sync their handheld devices to update their daily news and read the desired topics, and then they are able to discuss the most current issues during class time.

Using eAudio within the Curriculum

eAudiobooks allow students to read above their actual reading level. Therefore, a fifth grade teacher can download a collection of classical adventure novels in MP3 format onto her desktop. The students can then be assisted in selecting their favorite eBook and downloading it onto their handheld devices. Students can listen to their selected eAudiobook chapter-by-chapter and then summarize each chapter in a word processing program.

An 11th-grade teacher of literature can assign her students to read two of Shakespeare's works—one comedy and one tragedy. She can then download the entire works of Shakespeare in MP3 format onto a desktop computer. Students can then download the selected book and listen to them during out-of-school hours. While listening to the eAudiobook, students can record their reactions on the notepad or word processor on the handheld device. During class time, students can form literature circles to compare the emotional impact of each style of writing when they listen to and compare the same passage.

An eighth grade English teacher can use the features of voice synthesis software TextAloud 2.0 <www.nextup.com/> to convert text into spoken audio. Each eighth grader can write a short story using the basic rules of composition and creative writing. Instead of letting the project stop at that level, the teacher can then help the students convert the text of their story to MP3 format. Students can then upload their classmates' audio stories and listen to them on their mobile devices during out-of-school hours.

Mobile Devices for Students with Special Needs

Mobile devices have become a great asset for students with special needs. In the same school, the special needs teacher of a visually impaired student who was mainstreamed into the regular classroom was constantly on the lookout for materials similar to what was used in the classroom, but in a different format. She often used Bookshare.org.

Bookshare.org <www.bookshare.org> is an online community that enables people with visual and other print disabilities to share legally scanned books. Bookshare.org makes its materials available to U.S. residents who have a visual impairment, learning disability, or mobility impairment and to schools or groups that serve individuals with print disabilities.

Bookshare.org makes its materials available to users in BRF, DAISY, HTML, and text formats, which can be used with a browser to vary the text size or with a text-to-speech program. They provide a wide selection of eAudiobooks in MP3 format.

DAISY (Digital Accessible Information System) is a standard for producing books in an electronic format accessible for print-impaired readers. The DAISY standard offers people with learning disabilities a multisensory approach to reading. The cost for an annual subscription with no per-book download charges is $25 to sign up and $50 for the annual subscription. Bookshare.org's goal is to break even financially with modest membership fees and volunteer support from its community of members and supporters.

Using eVideo within the Curriculum

Because eVideo requires a large amount of storage space, its use in the classroom is just recently becoming commonly used. One way a middle school reading teacher of reluctant readers can use eVideo is to select a video with closed captioning or subtitle features. She can turn off the volume and turn on the captioning or subtitle feature. The visual cues will increase the student's recall of the printed word and the student will be better able to summarize the information in the video than if he had relied only on reading the printed text.

One American history teacher who made regular use of digital curriculum such as Discovery Education, which provides unitedstreaming <www.unitedstreaming.com> and Digital Curriculum <www.digitalcurriculum.com/>, often had trouble making those videos accessible to students who were absent. With a special grant, the teacher was able to purchase a handheld computer with a SecureDigital (SD) slot. He then downloaded the video onto the handheld device's SD memory card. When a student missed a video assignment, she would be able to check out the handheld device and watch the eVideo outside of classroom time.

In downloading eVideo onto handheld devices, care needs to be taken to consult the service's terms of use so that copyright laws are not violated. Many services permit their digital formats to be transferred to CD, DVD, or other physical media for student and teacher multimedia projects or student and teacher educational (and non-commercial) presentations only. These agreements broaden the educational usefulness as they allow students to download videos onto handheld devices for viewing at their convenience.

Using Podcasting within the Curriculum

Within a few months, podcasting has exploded from an experimental exercise for computer geeks to an educational tool for all levels and content areas. Podcasts can be cataloged and played back repeatedly. Because podcasting is an unregulated area and not governed by FCC rules, classroom teachers need to preview a podcast before selecting one for student use.

One middle school social studies teacher subscribed to a podcast produced by a middle school in Australia. From their program, the students are able to compare and contrast the similarities of the two cultures and educational systems. This class of eighth graders learned to listen to the differences in dialect and vocabulary and the diversity in language use, patterns, and social roles.

One senior class government teacher downloaded selected podcasts from a variety of political leaders, parties, and pressure groups. While listening to the podcasts, the students took notes about the particular stand the speaker took on an issue and the supporting facts he used to persuade his audience as to the reasons why his stand is the best way to approach a given situation.

A junior Spanish class subscribed to two separate podcasts—one from Mexico and one from Spain. The teacher then directed class discussion about the differences in the daily use of the same language.

Producing podcasts in the classroom can meet a multitude of cross-curriculum standards and benchmarks along with adding a great deal of value to the class work of the student. Many of these are covered in the book *KidCast: PodCasting in the Classroom* by Dan Schmit.

One ninth grade English teacher helped her students develop a weekly podcast in which they read the poetry they had authored and book talks they had given. These podcasts were produced as cross-curricular projects as they worked with the science department in learning the fundamentals of digital audio technologies including acoustics, digital sound compression, and networked distribution of content. They consulted with the mathematics department in calculating file size, compression, bandwidth, and server size.

As part of a study on Africa, a sixth grade class worked in groups of five. Each group researched a different region and presented their findings in a 10-minute presentation. These sessions were broadcast on the Internet via a podcasting site. Students conducted interviews and online research, wrote scripts, and selected accompanying music. They made decisions about how to present their information, whether it was lecture style or using interviews or a mix of the two. The broadcasts included recorded drumming sequences performed in class.

Summary and Challenge

Handheld devices have enormous potential for education. Educators around the world, regardless of the grade level or subject content, are amazed how engaged their students are when they are using handhelds. The result of "anytime/anywhere learning" is coming from the decreasing size of handheld devices, their availability, lower costs, and new types of wireless connectivity. With a little creativity, educators can take full advantage of this new format of learning and apply it to their classroom instruction.

Before continuing, ask yourself the following questions: "How could taking advantage of the smaller size and portability of handheld devices enhance your curriculum?" "How might the experiences of other educators using handheld devices in the classroom be adopted to your particular library or classroom situation?"

chapter thirteen

Record Keeping on Handheld Devices

Business and school leaders were first to take advantage of the administrative organization features of handheld computers. They were now able to access their personal schedules, notes, off-line email, and address book from a portable device. As users recognized the advantage of having organizational power in the palm of their hand, they began to search for other ways in which a handheld device could make their work easier.

Inventory Control

Inventory control is one of the primary uses of handheld devices within the school. Most circulation packages used by school libraries contain an add-on module that includes a barcode reader for inventory purposes. Examples of inventory modules for school libraries include the Sagebrush™ In-Hand handheld computer solution from Sagebrush Corporation <www.sagebrushcorp.com/tech/inhand.cfm> and the In-Path Portable Terminal Devices <www.inpath.com/pordatter.html> compatible with Follett Library Software.

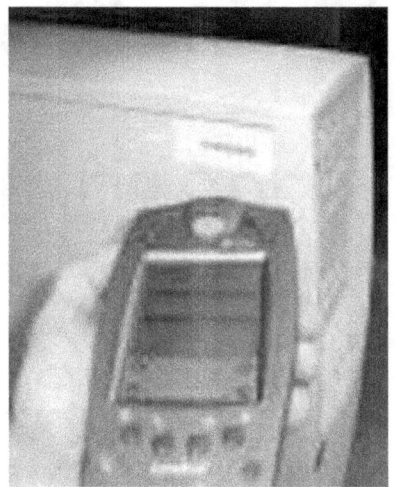

Figure 13:1 - Symbol Handheld Computer

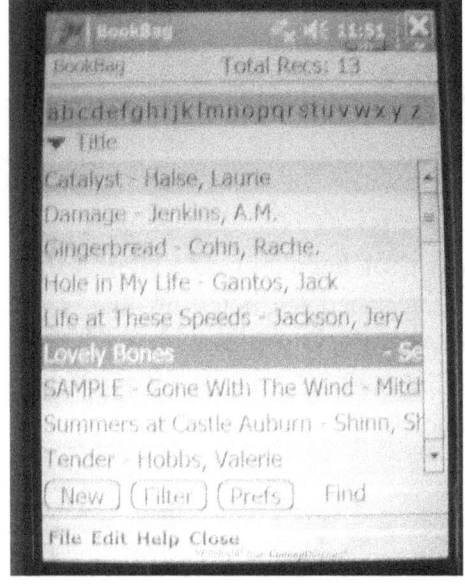

For smaller collections, such as classroom and personal libraries, BookBag™ <www.wakefieldsoft.com/bookbag/> organizational software is designed for the Palm OS handhelds, Pocket PC or Windows Mobile handhelds, and Windows PCs. For under $25, this software can track books, magazines, and other print materials that are owned, that have been read, that have been checked out, or that are wanted to read or buy.

Figure 13:2 – Screenshot of BookBag Software

AudioList <www.wakefieldsoft.com/audiolist/> is complete audio organization and inventory software for the Palm OS handhelds, Pocket PC or Windows Mobile handhelds, and Windows PCs. This software can track collections of podcasts, MP3s, tapes, CDs, and music that is owned, that is listened to, that is loaned to others, or that is desired to purchase.

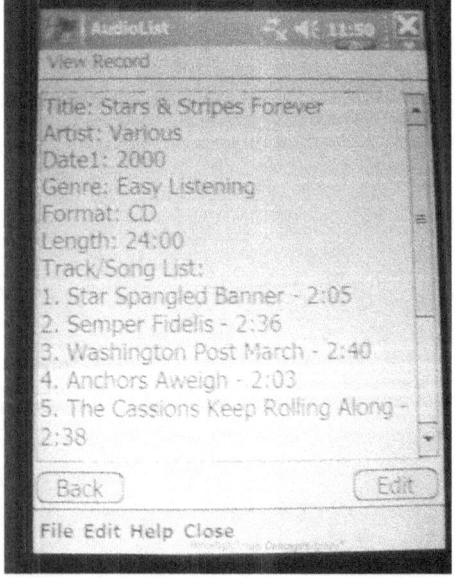

Figure 13:3 – Screenshot of AudioList Software

VideoList™ <www.wakefieldsoft.com/videolist/> is complete eVideo, movie, video, or DVD organization software for the Palm OS handhelds, Pocket PC or Windows Mobile handhelds, and Windows PCs.

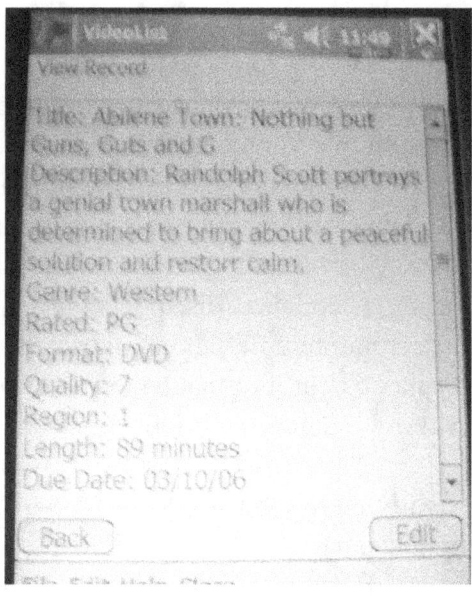

Figure 13:4 – Screenshot of VideoList Software

Handheld computers can also be used to enhance safety when used as a barcode reader for student identification or guest passes. Some programs are designed to deal with crisis or emergency situations. Tracker™ and Seeker™ <www.schoolid.com/> from Student Tracking Solutions lets schools quickly account for students in an emergency by having each teacher log in the students assigned to their supervision. The teachers and administrators then beam their information to a central source to create an overall school attendance file.

Schedule and Appointment Reminder

One of the most powerful applications on most handheld devices is the calendar program, which can be used to track a schedule and to remind the user of important tasks that need to be done. An audio alarm or reminder is available for use on Palm or Windows Mobile Operating Systems. Users can access their calendar on their desktop computer through Outlook or a similar scheduling program. They can even synchronize their handheld calendar to a Web-based calendar such as the one offered by Yahoo!.

Track Student Progress

Handhelds can be used to record student progress without interfering with the instructional process. Data can be captured in real time and then transmitted to the teacher's class file for compilation and analysis. Teachers can have an automated way to keep running records using a program such as mCLASS <www.wirelessgeneration.com/web/products.html>.

Many tools for tracking student progress on handheld computers are available. The simplest way is by making one's own templates that are set up in a spreadsheet or database program. Because the user sets up the template, it can be as simple or complex as desired. Conduct assessments often require immediate, accurate documentation. Therefore, educators can develop basic checklists, rubrics, and other authentic assessment tools using spreadsheets, databases, or even simple authoring tools.

Learner Profile <www.learnerprofile.com/> is an excellent assessment management tool to track Adequate Yearly Progress. Learner Profile gives teachers the ability to record students' grades, track assignments, organize student information, and develop reports.

Recording and Tabulating Grades

Many teachers like keeping grades on their handhelds wherever they go and having grades available to be tabulated, calculated, graphed, charted, and analyzed with no extra effort. Grading tools range from simple "homemade" solutions to comprehensive district-wide systems. Many systems allow teachers to import student data from their student information system, so they do not have to re-key the data. One popular student information system with a handheld module is the Star_Base School Suite <www.centuryltd.com/>. The handheld modules provide a broad range of features including: busing, attendance, discipline, grades, schedules, medical, student information, student search, access data from multiple schools, advanced query to retrieve students' records, barcode scan in attendance, display parent information, email, display emergency information, display events, print, scan in data, security features, staff lookup, student photos, and teacher schedules.

The most basic method to track grades is to keep a simple, teacher-created spreadsheet. A slightly more sophisticated way to track grades is to create a database that tracks student assignments. Free database templates designed specifically for education are available online.

Popular standalone grading programs include Excelsior Pinnacle <www.gradebook.com/> and Making the Grade <www.gradebusters.com/>.

Attendance Tracker

Examples of inexpensive software packages include Private Instructor's Log PPC <www.baggetta.com/privateinstructorlog.htm>, a database program designed for independent instructors or teachers of just about any content or skill area. This software tracks student personal information, billing, lessons, grades, and other routine tasks. It includes a note pad for each lesson to keep track of assignments, parental contacts, and instructions.

AttendanceTrackerPPC Ver 3.0 <www.baggetta.com/attendtracker.htm> only does one thing, but it does it efficiently. This program helps the teacher become more efficient at the job of tracking attendance—absenteeism, tardiness, skipping, classroom excuses, and disciplinary removals—all at the click of the screen.

Pocket PassTracker Logger <www.baggetta.com/passtrackerlogger.htm> is a small, helpful database program for anyone who has to keep track of the whereabouts of students, employees, or club members. It was designed primarily for teachers who write a lot of passes and have to log the information for administrative purposes, but it has a multitude of uses in education, business, health fields, and hobbies.

Lesson Planner

Lesson plans stored on handheld devices are easily accessible and searchable. These lesson plans can be in a handheld word processor, outliner, or database or with third party software such as Tapperware <www.tapperware.com/>. Another helpful software program is the Portable Lesson Planner PPC (for Pocket PC) <www.baggetta.com/portableplanner.htm>, which helps teachers keep track of where they are going in the curriculum. With this inexpensive software, teachers are able to check, create, and update their lesson plans on the fly.

Student Information Systems (SIS)
All schools have student information systems (SIS) with databases of student information, including demographic information, schedules, medical data, emergency contact information, discipline records, grades, and attendance. Handheld devices provide instant access to that data.

Some popular applications that provide instant access to student data include Tracker and Seeker <www.schoolid.com/>, ePrincipal® <www.media-x.com/products/eprincipal>, Austin Sky Technology offers Teacher Evaluation Software and Student Data Mobile <http://www.austinsky.com/>, and MarkBook® <www.markbook.com>.

Student Organizational Software
Not only do teachers need organizational software for handheld devices, but students also can benefit from software designed specifically for their needs. For $19 the software 4.0Student™ <www.handmark.com/products/detail.php?id=46&r_id=05_aug15_plm_nsltr> is the premiere coursework, class information, and grade tracking software for Palm OS devices. 4.0Student contains the most feature-rich and robust coursework management functionality of any student software available for the Palm OS computing platform.

ClassWork Organizer <http://downloads-zdnet.com.com/3000-2161-10147721.html> is designed to help students improve their educational productivity with a simple and attractive interface for organizing their schoolwork and activities. This software can display schoolwork entries in chronological order by due dates in an appropriate and organized manner.

Managing Classroom Sets of Handheld Computers
Managing classroom sets of handheld computers can seem overwhelming at times. For approximately $500 GSS HiHo–Classroom Edition <www.grantstreetsoftware.com/Products/gsshiho/index.html> empowers educators to integrate handheld technology into the classroom by providing a simple and effective method for handing out documents to students and allowing students to hand-in documents to their teacher. GSS HiHo–Classroom Edition helps bridge the document-transfer gap existing between students, their handhelds, and teachers in the classroom. When GSS HiHo–Classroom Edition is installed on a desktop computer, it creates a Hand-In and Handout folder, which provides a central point from which the teacher can distribute and receive files from students. As students perform synchronization operations using the classroom computer, the GSS HiHo–Classroom Edition HotSync Conduit allows students to hand in documents to the teacher and hands out documents to the student. This software can extend the capability to a greater number of students and devices, allowing several classroom computers as HotSync stations.

The management techniques of handheld devices within a classroom setting are dependent on the number of handhelds and the objectives of the teacher. With the one-to-one arrangement, each student has his or her own handheld. The teacher must decide if a student has the handheld device for the entire school year, a selected period, only for a project, or only during the school day.

The 24/7 model permits students to have the handheld devices 24 hours a day, seven days a week. Students have sole possession of the computer for a school year, a semester, or the time of a project.

The one-to-two model allows pairs of students to share one handheld. Teachers decide how students will store their files and how the handhelds will be named. This model lends itself more toward small-group instruction, since the whole class cannot be involved in the same handheld activity.

Many schools purchase a class set of handheld devices that must be shared among many teachers. Sharing classroom sets can be challenging to manage because many students must use the same handheld. In these situations, it is helpful for each student to use the same handheld from the set. There will also be issues with students from other classes viewing, modifying, or deleting other students' data. Because handheld computers do not have the security features that desktop computers have, this could be the greatest challenge in using the class set checkout model.

In some cases, teachers may have as many as eight or as few as one handheld device to use with an entire class. This arrangement lends itself to individual and small group work. Only selected or volunteer students use the selected handhelds, and the teacher needs to decide how to protect files from other students.

Besides labeling each handheld, a handheld device also needs a unique user name in its memory. The user name is assigned the first the handheld is synchronized with the desktop or with some systems when the desktop software is installed. If students have their own handheld, the handheld can be named after the user. If more than one student is using the same handheld, teachers may wish to choose a sequence of numbers or letters for the name of the handheld.

Once the handheld is assigned a name, the desktop computer creates a user folder with this name during synchronizing. When a user synchronizes that handheld, all of his or her data and work is stored in this folder. The desktop software makes it easy to switch between users, so having multiple student folders on the same computer is not a problem.

The original Palm Pilots ran on AAA batteries. Teachers and students found themselves continually buying and charging batteries. Today, almost all handhelds come with built-in rechargeable batteries. Many teachers set up banks of chargers on power strips so students can be responsible for keeping their own handheld device charged. One possibility is the Portsmith Multi-slot Charging Cradle that provides simultaneous charging for up to 10 handhelds at a time.

Managing Computer Networks with Handheld Devices

From a central location, information technology (IT) administrators can solve mobile management challenges and streamline systems management with programs such as Afaria from iAnywhere® <www.ianywhere.com/products/afaria.html>. Users can manage a broad range of frontline devices, such as laptops and tablets, handhelds, Pocket PCs, smartphones, POS systems, and remote desktops and servers. They can capture and store hardware and software information, and automatically keep track of mobile devices and their health. Afaria can also assist in providing automated system and data backup and restore capability in the event that data is destroyed or lost.

SonicAdmin from Avocent® <www.sonicmobility.com/> helps IT personnel manage the network and server environment from their handheld computers. With this software, users can manage users, event logs, DNS, printers, files, services, processes, and more. It also includes the ability to talk to any server or network device at the command-line level.

Summary and Challenge

With the increasing demands on educators and reducing funding, it becomes necessary for teachers and school media specialists to become more creative in their search for additional tools to help their students meet the necessary standards and benchmarks. At the same time, teachers have more paperwork and documentation required of them. Whether it is taking inventory, recording grades, attendance, lesson plans, or notes, there are several software packages for handheld devices available for any need. Every minute saved using handheld devices can be minutes applied to direct student instruction and preparation. Even students can better organize their school, work, or personal lives with handheld devices.

Before continuing ask yourself these questions: "What routine classroom or library tasks could be simplified with the use of a handheld device?" "What software would best fulfill that need?" "How could you encourage students to use handheld devices as personal organizers?"

appendix A

eBook, eAudio, and eVideo Formats

.aa — audible audiobook format

.AAC — Advanced Audio Coding from QuickTime. It was popularized by Apple Computer through its iPod and iTunes Music Store

.aiff (AIFF) — Audio Interchange File Format for Macintosh Computers

.asf (ASF) — Advanced Systems Format (ASF) is the file format used by Windows Media. Audio and/or Video content compression

.avi (AVI) — Audio Video Interleave. Windows video format

.bmp (BMP) — standard bit-mapped graphics format used in the Windows environment

.btf (BTF) — Braille ready file for use with a refreshable Braille display or Braille embosser

.divx (DivX) — digital video express, a DVD-ROM format

.doc —Microsoft Word document file

.ebo (EBO) —MS Reader annotations file

.fla (FLA) — Macromedia Flash Authoring file

.gif (GIF) — image format designed for drawings

.html/xml — HTML eBook

.jpeg (JPEG) — image format designed for photographs

.lit (LIT) — Microsoft Reader format

.mov (MOV) — QuickTime video

.mpeg (MPEG-1) — coding of moving pictures and associated audio for digital storage media at up to about 1.5M bit/s

.mp2 (MPEG-2) — used in digital TVs, DVD-Videos, and in SVCDs

.MP3 (MPEG-3) — MP3 stands for MPEG-1 Audio Layer III

.mp4 (MPEG4) — digital video codec standard, which is noted for achieving very high data compression

.mpg (MPEG) — video format

.ogg (Ogg Vorbis) — new audio compression format

.pdb (PDB) — eReader and eReader Pro for Palm OS

.pdf (PDF) — Adobe Acrobat Format

.php (PHP) — an HTML-embedded scripting language for authoring Web pages

.pml (PML) — Palm Markup Language
.png (PNG) — image format for photographs or drawings
.pps (PPS) — PowerPoint show file
.ppt (PPT) — PowerPoint presentation file
.prc (PRC) — MobiPocket Reader
.rm (RM) — RealMedia audio/video
.rtf (RTF) — rich text format
.rwp (RWP) — ReaderWorks project file
.txt (TXT) — Plain Text
.wma (WMA) — Windows Media Audio
.wmv (WMV) Windows Media Video
AVC — Advanced Video Coding
H.264 — digital video codec standard with very high data compression
M4A — M4A is the new replacement for the older MP3 audio format
MIDI — Musical Instrument Digital Interface

sources consulted

About AudioFile. 26 Feb. 2006 <www.audiofilemagazine.com/?about.html>.

American Association of School Librarians, and Association for Educational Communications and Technology. *Information Literacy Standards for Student Learning*. 1998. American Library Association. 2 Aug. 2006 <http://www.ala.org/ala/aasl/aaslproftools/informationpower/InformationLiteracyStandards_final.pdf>.

Arnsdorff, Marvin, Dr. *Mounting Research on Backpack Use*. May-June 2002. International Chiropractic Pediatric Association. 26 Feb. 2006 <http://www.icpa4kids.org/research/articles/childhood/backpack_research_newsletter.htm>.

Bailey, Gerald D., and David Pownell. *Administrative Solutions for Handheld Technology in Schools*. Eugene, OR: International Society for Technology in Education (ISTE), 2003.

"Best Sites for Windows Powered Pocket PCs, Handheld PCs, and Smartphones." *Smartphone & PocketPC Magazine*. 26 Feb. 2006, 1 Aug. 2006 <www.pocketpcmag.com/?_top/?bestsites.asp>.

"Beyond the Text: Comparison Chart of e-Book and Digital Talking Book (DTB) Hardware and Software." *National Center for Assessible Media - Projects*. National Center for Assessible Media. 26 Feb. 2006 <http://ncam.wgbh.org/ebooks/comparison.html>.

Brazell, Wayne. "Handheld Computers: A Boon for Principals." *Principal-Politics and the Principalship-Tech Support* Jan.-Feb. 2005: 48. National Association for Elementary School Principals. 26 Feb. 2006 <www.naesp.org/?ContentLoad.do?contentId=1461>.

Breen, Christopher. "How to Create a Vodcast: Steps for Offering Video on Demand." *Playlist* 26 July 2005. 26 Feb. 2006 <http://playlistmag.com/?features/?2005/?07/?howtovodcast/?index.php>.

Bright, Phoebe. "Podcasting in Schools - Research." *Bringing Clarity to Change*. 23 Feb. 2005. Vivid Logic. 25 Feb. 2006 <www.vividlogic.ie/?index.php/?vividlogic/?more/?podcasting_and_schools_research>.

Brumfield, Robert. "Study: States Are Slowly Embracing eTexts." *eSchool News Online* 28 Sept. 2005. 26 Feb. 2006 <www.eschoolnews.com/?news/?showStory.cfm?ArticleID=5883>.

Carr, Nora. "Try 'Podcasting' to Broaden Your PR Reach." *eSchool News* Sept. 2005: 39.

"CAST: Universal Design for Learning." *CAST: Center for Applied Special Technology*. 26 Feb. 2006 <www.cast.org/>.

Caughlin, Janet. *Handhelds for Teachers & Administrators: A Complete Resource for Using Handhelds in Grades K-12!* Watertown, MA: Tom Snyder Productions, 2003.

Cavanaugh, Terence W. *The Digital Reader: Using E-Books in K-12 Education.* Eugene, OR: International Society for Technology in Education (ISTE), 2006.

- - -. "eBooks and Reading: eBook Reading Strategies (eBRS)." *eBRS.* 20 May 2006. 2 Aug. 2006 <www.drscavanaugh.org/?ebooks/?ebrs/?intro.htm>.

Cha, Bonnie. "Talk to the Hand(held): PDAs with Voice Recorders." *ZDNet: Reviews.* CNET Networks. 13 Sept. 2005. 2 Aug. 2006 <http://reviews-zdnet.com.com/?4520-3127_16-6322590-1.html?tag=nl.e539>.

Church, Audrey R. "E-Book Resources for the School Library." *MultiMedia@Schools* July-Aug. 2005: 9-12.

Cochrane, Todd. *Podcasting: The Do-It-Yourself Guide.* Indianapolis: Wiley Publishing, 2005.

Copyright Condensed. Johnston, IA: Heartland AEA, 1999.

Crawford, Valerie, and Phil Vahey. "March 2002 Evaluation Report." *Palm Educations Pioneer Report* Mar. 2002. 26 Feb. 2006 <http://palmgrants.sri.com/?PEP_R2_Report.pdf>.

Crews, Kenneth D., Ph.D. "New Copyright Law for Distance Education: The Meaning and Importance of the TEACH Act." *Distance Education and the TEACH Act.* 30 Sept. 2002. American Library Association. 25 Feb. 2006 <www.ala.org/?ala/?washoff/?WOissues/?copyrightb/?distanceed/?distanceeducation.htm#newc>.

Curtis, Michael, et al. *Palm Handheld Computers: A Complete Resource for Classroom Teachers.* Eugene, OR: International Society for Technology in Education, 2003.

DAISY Consortium Web Site. 17 July 2006. 2 Aug. 2006 <www.daisy.org/>.

"E-book Readers Comparison Chart." *eBooks from Harper Collins.* Perfect Bound. 26 Feb. 2006 <http://us.perfectbound.com/?AACDDF66-9D84-4ED8-84C9-F053B8B4C3BC/?10/?1/?en/?Help-Reader-General.htm>.

"Educational Technology Standards and Performance Indicators for Administrators." *National Educational Technology Standards for Administrators.* International Society for Technology in Education. 25 Feb. 2006 <http://cnets.iste.org/?administrators/?a_stands.html>.

Fair Use Guidelines for Educational Multimedia. 17 July 1996. The Consortium of College and University Media Centers ("CCUMC"). 26 Feb. 2006 <www.ccumc.org/?copyright/?ccguides.html>.

Fasimpaur, Karen. *101 Great Educational Uses for Your Handheld Computer: A Comprehensive Guide to Using Handhelds in Education for Administration, Teaching, and Learning.* Long Beach, CA: K12 Handhelds, 2003.

Fox, Steve. "The Podcasts Are Coming." *PC World* Oct. 2005: 26.

Gade, Lisa, Editor. "Palm vs. Pocket PC: Which One Is for You?" *Mobile Tech Review* June 2004. 26 Feb. 2006 <www.mobiletechreview.com/?tips/?palm_vs_pocketpc.htm>.

Google Book Search Library Project. Google. 26 Feb. 2006 <http://print.google.com/?googleprint/?library.html>.

Hall, Rich, Derek Ball, and Barry Shilmover. *How to Do Everything with Your Dell Axim Handheld.* Emeryville, CA: McGraw-Hill/Osborne, 2006.

"Handheld Computers." *Becta.* July-Aug. 2003. 26 Feb. 2006 <www.becta.org.uk/?subsections/?foi/?documents/?technology_and_education_research/?handheld_computers.pdf>.

Hoffman, Gretchen McCord. *Copyright in Cyberspace: Questions and Answers for Librarians*. New York: Neal-Schuman, 2001.

Hungerford, William. "Palm vs. Pocket PC-The Great Debate? Is There a Right Choice?" *About.com* 26 Feb. 2006 <http://palmtops.about.com/?cs/?pdafacts/?a/?Palm_Pocket_PC.htm>.

Information Technology. Dept. home page. 4 Feb. 2005. Northwestern University. 26 Feb. 2006 <www.it.northwestern.edu/?hardware/?pda/?standard.html>.

Johnson, Doug. "E-Books, E-Learning, E-Gads! How Librarians Can Avoid Extinction." *School Library Journal*. Nov. 2004. 26 Feb. 2006 <www.doug-johnson.com/?handouts/?ebook.pdf>.

Karpen, Jim, Ph.D. "Downloading Web Sites to Your Pocket PC for Offline Reading." *PC Magazine* July 2003. 26 Feb. 2006 <www.pocketpcmag.com/?_archives/?jul03/?online.asp>.

Listening Factoids. 20 Sept. 2004. International Listening Association. 26 Feb. 2006 <www.listen.org/?pages/?factoids.html>.

Lutzker, Arnold P., Esq. "What the Digital Millennium Copyright Act and the Copyright Term Extension Act Mean for the Library Community." *Primer on the Digital Millennium* 8 Mar. 1999. Association of Research Libraries. 25 Feb. 2006 <www.arl.org/?info/?frn/?copy/?primer.html>.

Magid, Larry. "DIY: Create Your Own Podcasts." *PC Magazine* 25 Sept. 2005. 26 Feb. 2006 <www.pcmag.com/?article2/?0, 1895, 1863364, 00.asp>.

McElhearn, Kirk. "Ripping Audiobooks." *PC Magazine* 17 Aug. 2005. 26 Feb. 2006 <www.pcmag.com/?article2/?0%2C1895%2C1846814%2C00.asp>.

"Microsoft Releases Windows Mobile 5.0." *Press Pass Information for Journalists* 10 May 2005. 26 Feb. 2006 <www.microsoft.com/?presspass/?press/?2005/?may05/?05-10WindowsMobile5PR.mspx>.

Mid-Illinois Talking Book Center. 26 Feb. 2006 <www.mitbc.org/>.

"NETS for Students: Technology Foundation Standards for All Students." *ISTE National Education Technology Standards for Students*. International Education Technology Standards for Students. 25 Feb. 2006 <http://cnets.iste.org/?docs/?NETS_S.doc>.

"NETS for Teachers: Educational Technology Standards and Performance Indicators for All Teachers." *ISTE National Educational Technology Standards for Teachers*. International Society for Technology in Education. 25 Feb. 2006 <http://cnets.iste.org/?docs/?NETS_T.doc>.

Paar, Morgan. "Video 2Go." *Videomaker* Mar. 2006: 56-58.

Pascopella, Angela. "Digital Days." *District Administration Magazine* Aug. 2005. Professional Media Group. 26 Feb. 2006 <www.districtadministration.com/?page.cfm?p=1189>.

Pash, Adam. "MOTO ROKR - Motorola Introduces the First iTunes Phone - The ROKR." *About.com* 26 Feb. 2006 <http://mp3.about.com/?od/?mp3players/?a/?motorokrpeak.htm>.

"Placing Annals Articles or Images on Your PDA Using Web Clipping Software or Mobile Favorites." *Annuals of Internal Medicine*. American College of Physicians. 26 Feb. 2006 <www.annals.org/?pda/?webclippers.shtml>.

Podcast Creation Guide. Apple Computers, 2005. 26 Feb. 2006 <http://images.apple.com/?education/?solutions/?podcasting/?pdf/?PodcastCreationGuide.pdf>.

"Podcasting." *Podcasting- K-12 Handhelds*. K-12 Handhelds. 25 Feb. 2006 <www.k12handhelds.com/?podcasting.php>.

Pownell, David, and Gerald D. Bailey. *Administrative Solutions for Handheld Technology in Schools*. Eugene, OR: International Society for Technology in Education (ISTE), 2003.

Price, Gary. "Search and Read Full Text Books Online via ebrary." *Search Engine Watch* 14 Nov. 2005. 26 Feb. 2006 <http://blog.searchenginewatch.com/?blog/?051114-110116>.

Rosenborg, Victoria. *EPublishing for Dummies*. Foster City, CA: IDG Books, 2001.

Schleicher, Stephen. "Recording Your Show." *Podcasting 102* 14 Sept. 2005. Corporate Media News. 26 Feb. 2006 <www.corporatemedianews.com/?articles/?viewarticle.jsp?id=34618>.

Schmit, Dan. *KidCast: Podcasting in the Classroom*. N.p.: FTC Publishing, 2005.

Simpson, Carol. *Copyright for Schools: A Practical Guide*. 4th ed. Worthington, OH: Linworth Publishing, 2005.

"Specifications for the Digital Talking Book." *An American National Standard Developed by the National Information Standards Organization* 6 Mar. 2002. American National Standards Institute. 26 Feb. 2006 <www.niso.org/?standards/?resources/?Z39-86-2002.html>.

Standards for the English Language Arts. National Council for Teachers of English. 26 Feb. 2006 <www.ncte.org/?about/?over/?standards/?110846.htm>.

Sullivan, Laurie. "Finding Profits in Podcasting." *Information Week* 29 Aug. 2005: 47, 49-50.

U.S. Department of Education. Office of Educational Technology. *National Education Technology Plan 2004*. Washington, D.C., 2004. 5 Aug. 2006 <http://www.ed.gov/about/offices/list/os/technology/plan/2004/site/edlite-background.html>.

"Vodcast." *Wikipedia* 23 Feb. 2006. 25 Feb. 2006 <http://en.wikipedia.org/?wiki/?Vodcast>.

"WeBlogs for Education." *World of Media* Oct. 2005: 2-4. Heartland Area Education Agency 11. 25 Feb. 2006 <www.aea11.k12.ia.us/?womwww/?05-06wom/?WOMOct2005.pdf>.

"Welcome to the IDPF." *International Digital Publishing Forum* 2 Feb. 2006. International Trade and Standards Organization for the Digital Publishing Industry. 26 Feb. 2006 <www.idpf.org/>.

Windows Media Player 10 Mobile. Microsoft. 26 Feb. 2006 <www.microsoft.com/?windows/?windowsmedia/?player/?windowsmobile/?default.aspx>.

"Windows Mobile Tips." *PC Today* Dec. 2005: 50-51. 26 Feb. 2006 <www.pctoday.com/?Editorial/?article.asp?article=articles/?2005/?t0312/?22t12/?22t12.asp&guid>.

glossary

ActiveSync - Microsoft's synchronization software for Windows Mobile-based devices. The software exchanges and updates information between a handheld and a desktop computer.

aggregator - a type of software that retrieves syndicated Web content that is supplied in the form of a Web feed (RSS, Atom, and other XML formats), and that are published by Web logs, podcasts, vlogs, and mainstream mass media Web sites.

beam, beaming - the act of transferring information from one infrared-enabled device to another. Infrared is commonly used in handheld computers and remote controls.

BlackBerry - a line of mobile email devices and services from Research in Motion (RIM).

blog - short for Web log. A blog is a Web page that serves as a publicly accessible personal journal for an individual.

Bluetooth - a short-range radio technology aimed at simplifying communications among Internet devices and between devices and the Internet. It also aims to simplify data synchronization between Internet devices and other computers.

ClearType - a software technology developed by Microsoft that improves the readability of text on existing LCDs (Liquid Crystal Displays), such as laptop screens, Pocket PC screens, and flat panel monitors.

CompactFlash (CF) - a small, removable mass storage device created by SanDisk. Cards weigh a half ounce and are the size of a matchbook.

conduit - software that enables information to be updated and exchanged between a desktop computer and a handheld device.

cradle - the stand that a handheld computer sits in to synchronize to a desktop computer; for rechargeable handhelds, the cradle also charges the device.

DAISY - acronym for Digital Accessible Information SYstem. A digital standard that permits everyone, but especially people who are blind, visually impaired, or have another print disability, to experience a better way to read.

eAudio - an electronic version of an audio file.

eBook - an electronic version of a book.

eVideo - an electronic version of a video.

expansion slot - a slot built into most handhelds that allows a memory card to be inserted and accessed by the handheld. The expansion slot also acts as a connection port for peripheral devices.

Flash - a bandwidth friendly and browser independent vector-graphic animation technology owned by Adobe Systems, Inc. (formerly Macromedia, Inc.).

Global System for Mobile Communications (GSM) - a digital cellular phone technology that is used worldwide.

GPRS - short for *General Packet Radio Service*, a standard for wireless communications that runs at speeds up to 115 kilobits per second, compared with current GSM (Global System for Mobile Communications) systems' 9.6 kilobits.

Graffiti - a software program from Palm, Inc. that enables a handheld device to recognize a proprietary handwriting language.

handwriting recognition software - software that can interpret handwritten text and convert it to digital text.

HotSync - the process that automatically exchanges and updates information between a Palm handheld device and the desktop software. After the HotSync, any changes that were made on the handheld or desktop will appear in both locations.

I/O ports - input/output ports between the Central Processing Unit and I/O devices in a computer.

ID3 Tag - a tagging format for MP3 files. It allows metadata such as the title, artist, album, or track number, to be stored in the MP3 file.

IEEE 802.11 - family of wireless connectivity technology developed by the IEEE (Institute of Electrical and Electronics Engineers).

infrared (IR) - wireless communication that uses light waves to transmit data.

Intel StrongARM Processor - the StrongARM microprocessor is a faster version of the Advanced RISC Machines ARM design.

Janus - is the codename for the portable version of Windows Media DRM for portable devices.

mAh - a milliampere-hour (abbreviated as mA·h) is a unit of electrical charge. It is a common measurement of how long a battery will last (or in the case of a rechargeable battery, how long it will last between charges).

miniSD card - a card from Sandisk Corporation that offers all of the benefits of a standard SD card, but is over 60 percent smaller than a full-sized SD card.

MP3 - A format for audio files that makes the file size small while preserving CD-quality audio.

MultiMediaCard (MMC) - A tiny memory card from Sandisk Corporation that offers portable storage and can be swapped between devices such as handhelds, mobile phones, and digital cameras.

NISO Z39.86 - standard that defines the format and content of the electronic file set that comprises a digital talking book (DTB) and establishes a limited set of requirements for DTB playback devices.

operating system (OS) - software that manages the overall operation of a computer system, including the allocation of storage and memory and scheduling tasks. Palm OS and Windows Mobile are some of the current operating systems for handhelds.

optical character recognition (OCR) - the recognition of printed or written text characters by a computer. This involves photoscanning of the text character-by-character, analysis of the scanned-in image, and then translation of the character image into character codes, such as ASCII, commonly used in data processing.

Palm OS - the operating system used on Palm handhelds and third-part devices such as the AlphSmart® Dana and the Sony CLIÉ®.

peripheral - a hardware device that can be added to, attached to, or used with a handheld to give it additional functionality.

Personal Information Manager (PIM) - a specific type of handheld device that serves the purpose of organizing personal information: PIMs may include calendars, address books, notepads, and calculators.

Pocket PC - a handheld-sized computer that runs the Windows Mobile operating system.

podcast - a Web feed of audio or video files placed on the Internet for anyone to download or subscribe to, and the content of that feed.

podcasting - the preparation and distribution of audio (and possibly other media) files for download to digital music or multimedia players, such as the iPod. A podcast can be easily created from a digital audio file. The podcaster first saves the file as an MP3 and then uploads it to the Web site of a service provider. The MP3 file gets its own URL, which is inserted into an RSS XML document as an *enclosure* within an XML tag.

podcatcher - Software that automatically downloads and aggregates podcasts for synchronization to the user's portable media player.

probes - devices that are made to measure scientific phenomena such as temperature or PH. Interfaces connect handheld computers and probes.

processor - the part of a computer (a microprocessor chip) that interprets instructions and does most of the data processing.

QVGA (Quarter Video Graphics Array) - (also known as Quarter VGA or QVGA) - the image size of 320 x 240 pixels size/resolution of a display. VGA is 640 x 480 pixels, thus QVGA is one-quarter the area.

RAM (Random-access memory) - the most common type of computer memory that can access data randomly without touching the preceding bytes.

resolution - refers to the number of pixels a screen can display. The higher the resolution, the better the image quality.

ROKR - The Motorola ROKR (pronounced "rocker") is the first cellular phone that includes Apple Computer's iTunes music player.

RSS Feed - an acronym for Real Simple Syndication or Rich Site Summary, a family of Web feed formats, specified in XML format used for syndicating Web content.

RSS Validator - a program that will catch common XML errors such as unescaped ampersands and high-bit characters; domain-specific errors such as missing required elements; and more subtle errors such as improper language codes in the <language> element.

SecureDigital (SD) - a type of storage device from SanDisk Corporation that allows users to transfer data between a handheld and add-on devices.

Skype - a proprietary peer-to-peer Internet telephony (VoIP) network.

smartphones - handheld device that integrates mobile phone capabilities with the more common features of a handheld computer.

Sony Memory Stick - a removable flash memory card format launched by Sony in October 1998.

stylus - a writing device that comes with handheld computers; it writes on the screen and looks similar to a pen.

Symbian OS - an operating system designed for mobile devices. It is the global industry standard operating system for smartphones with associated libraries, user interface frameworks, and reference implementations of common tools produced by Symbian Ltd.

sync - short for synchronizing.

synchronize - exchange and update information between a handheld and a desktop computer. Also known as ActiveSync (Windows Mobile-based handhelds) and HotSync (Palm-based handhelds).

Text-to-Speech (TTS) - the artificial production of human speech using software or hardware.

Transcriber - a software program that enables Pocket PC devices to recognize words written in cursive and converts to typed text.

Universal Serial Bus (USB) - a type of connection to a desktop computer that can be used to synchronize data. It is usually much faster than a serial connection.

VGA resolutions - 480 x 640 pixels.

vidcast - a video podcast.

vodcast - the online delivery of video on demand content via Atom or RSS enclosures.

Web clipping - a way in which news or information from wireless-ready Web sites is "clipped" from the site and delivered in a format optimized for viewing on the handheld screen.

Wi-Fi (Wireless Fidelity) - a brand owned by Wi-Fi Alliance that enables wireless Internet access. It is often used generically when referring to any type of 802.11 network developed by the IEEE (Institute of Electrical and Electronics Engineers) for wireless LAN technology.

index

A

Adobe Acrobat Format (.pdf) – 22, 43, 80, 125, 128
Adobe Content server – 34, 119
Adobe Photoshop Album – 34
Afternoon, a Story – 32
Alex Catalogue of Electronic Texts – 37
Amazon.com – 38
American Library Association – 63, 121, 122
Arnsdorff, Dr. Marvin – 33, 121
ASCAP (The American Society of Composers, Authors, and Publishing) – 74
Audacity – 64, 80, 119
Audible – 64, 80, 119
Audio Standards – 83
Audiobooks – xii, 24, 57, 59, 61–66, 68, 80, 89, 123
AudioFile – 62, 121
AudioList 63, 112
AvantGo – 46, 107
Axim – 29, 122

B

Backpacks – 33
Baen Free Library – 38
Bartleby Bookstore – 37
Berne Convention – 98
Bibliomania – 38
Blackberry – 14, 125
Blind and Dyslexic – 67
Block Recognizer – 16
blog – 46, 70, 71, 78, 83, 85, 88, 94, 95, 124, 125
BlogMatrix – 79
Bluetooth – 16, 19, 20, 22, 41, 56, 125
BookBag – 112
Bookshare.org – 39, 108

C

Cavanaugh, Dr. Terry – 104, 105, 122
Center for Applied Special Technology, (CASTCenter) – 23, 121
Chapura Pocket Mirror – 14, 22
CIA Publication Library – 38
ClearType – 35, 125
Compact Flash (CF) – 17–20, 125
copyright – 9, 15, 32, 38, 39, 52, 64–66, 73, 74, 82, 83, 89, 97–100, 109, 122–124
Copyright for Schools: A Practical Guide – 100, 122
Creative Commons License – 74
Curry, Adam 70, 71
Curriculum-on-Demand – 95

D

DAISY (Digital Accessible Information System) – 24, 39, 108, 122, 125
Dell – 29, 122
Digital Book Index – 37, 40
digital cameras – 19, 126
Digital Millennium Copyright Act – 99, 126
Digital Performance Right in Sound Recordings Act – 99
Digital Rights Management (DRM) – 24, 27, 34, 57
Digital Talking Books – DTB – 23, 24
Disabilities – 23, 24, 39, 40, 108
Documents To Go – 22, 104
DropBook – 25, 52

E

eAudiobooks – xii, 8, 11, 15, 55–60, 62, 67, 107, 1088
Ebook Reader – 15, 24, 25, 30, 34, 35, 50, 56
eBook Studio – 37
eBooks – xi, xii, xiii, 1, 6, 8, 10, 11, 13–17, 23–26, 30–40, 49, 50–60, 64, 66–68, 103–108, 119, 122, 125
eBooks.com – 37
eBRS Program – 104
Education Podcast Network – 71
eReader (.pdb) – 25, 36, 52, 119
EuroCool – 30
eVideos – xi, sii, xiii, 1, 5, 8, 10, 11, 13–15, 17, 30, 87, 88, 95, 108, 109, 113, 125
Excel – 22, 35, 53
expansion cards – 17, 20, 26
expansion slots – 17

F

Fair Use – xiii, 38, 98–100, 122
Fair Use Guidelines for Educational Multimedia – 99, 1222
Fictionwise eBooks – 37, 57, 64
Flash Player – 28, 91
Follette – ix, 37, 112

G

Gale Virtual Reference Library – 58
GarageBand – 74, 79, 82, 83
Garfield, Steve – 88
General Packet Radio Service (GPRS) – 19, 126
Global System for Mobile Communications (GSM) – 20, 126
Google – 38, 40, 42, 89, 122
GPS (global positioning systems) – 15, 18, 20
Graffiti – 16, 126

H

Handango – 30, 92
Harry Fox Agency, Inc. – 74
Hart, Michael – 36
Highly Interactive Classrooms, Curricula & Computing in Education (The Hi-Ce) – 30

HotSync – 22, 33, 115, 125, 128
HP – 29
HTML eBook (.html/xml) – 36, 50, 119
HyperCard – 32
hypertext fiction 32

I

I/O ports – 17, 126
ID3 tags – 65, 74, 85, 126
IEEE 802.11 – 20, 126
Information Literacy Standards – 9, 62, 121
InfoSpace – 42
infrared port (IRDA) – 17–19, 22, 56, 125, 126
In-Hand handheld computer – 112
In-Path Portable Terminal Devices – 112
International Children's Digital Library – 37
International Listening Association – 62, 123
International Reading Association – 2
International Society for Technology in Education (ISTE) – 4, 5, 7, 9, 121–123
Inventory Control – 112
iPaq – 29, 92
iPod – xi, xii, 15, 17, 38, 54, 59, 64–67, 69–71, 76, 79, 80, 82, 83, 85, 87 94, 119, 127
iPod eBook Maker – 54
iPod Nano – 15, 54
iSiloX – 45
iStory Creator – 54
iSync – 14
iTalk – 65, 66
iTunes – 14, 15, 64, 65, 69, 71, 76–79, 82, 83, 85, 88–91, 93, 94, 119, 123, 127
iTunes U – 71

K

K12 Handhelds – 30, 72, 122
Keyboards – 16, 17, 19, 20, 43
KidCast – 72, 109, 124
King, Stephen – 32

L

LearnOutLoud.com – 59, 64, 73, 89
LibriVox – 64
LibWise – 57

M

MAC OS X – 79, 82, 85, 93, 107
Manybooks.net – 38
MARC records – 58
media players – xii, 15, 26, 30, 76, 81, 90, 95, 127
memory – 14, 15, 17, 18, 20, 26, 28, 29, 35, 46, 79, 92, 93, 109, 116, 125–127
MemoWare – 37
microphone – 65, 66, 74, 75, 82, 84, 95
microprocessor – 17, 126, 127
Microsoft ActiveSync – 22
Microsoft Reader – 25, 35–37, 51, 52, 57, 67, 106, 119
Microsoft Word eBook (.doc) – 36
Mid-Illinois Talking Book Center – 67, 123
Mixers – 75
Mobile Favorites – xii, 43 – 45, 123
MobiPocket Reader (.prc) – 25, 34, 35, 46, 53, 120
MultiMedia Card (MMC) – 18, 29, 126

N

Namo HandStory Suite – 45
National Center for Accessible Media (NCAM) – 23
National Council of Teachers of English (NCTE) – 2
National Education Technology Plan – 3, 4, 124
National Educational Technology Standards for Students (NETS-S) – 4, 5, 123
National Educational Technology Standards for Teachers (NETS-T) – 6, 7, 123

National Information Standards Organization – 32, 124
National Institute of Standards and Technology (NIST) – 32
NetLibrary – 58, 59
NISO Z39.86 – 23, 32, 39, 126

O

OCLC (Online Computer Library Center) – 57
Odiogo – 78
Ogg Vorbis – 28, 68, 92, 119
Online Radio – 67
Open eBook Forum (OeBF) – 23, 32
optical character recognition (OCR) – 50, 126
OverDrive Technology – 59

P

Palm – xi, 14–17, 19, 22, 25, 26, 28, 30, 33–36, 41, 42, 45, 52, 53, 56, 57, 63, 92, 107, 111–113, 115, 116, 119, 120, 122, 123, 126, 126, 128
Palm Boulevard – 30
PalmGear – 30
Photo Story – 27
Plain Text (.txt) – 24, 34, 50, 54, 120
Playlist – 26–28, 81, 82, 94, 121
PlaylistMag.com – 64, 82
Pocket DVD – 29, 92, 93
Pocket Gear – 30
Pocket Informant – 22
Pocket PC – 14, 16, 17, 19, 22, 25–30, 34, 37, 38, 42, 45, 46, 52, 53, 56, 57, 59, 63, 67, 68, 77, 78, 80, 81, 90, 92 94, 107, 112–114, 116, 121–123, 125, 127, 128
Pocket PC City – 30
PocketDISH – 90
PocketGear – 29, 30, 46, 68, 78, 92, 93
Pocket Movies – 90
PocketMusic Player – 29
PocketStreamer – 29, 68, 92

PocketTV – 29, 90
podcast – xiii, 1, 5, 6, 8, 10, 11, 13,64, 69–88, 90, 94, 95, 98, 99, 103, 105, 106, 109, 112, 121, 122–125, 127, 128
Podcast Alley – 80, 85
Podcast Factory – 74
Podcasting in Education – 72
Podcatcher – 70, 76–78, 80, 127
PodSpider – 80
PowerPoint – 16, 22, 35, 53, 120
Processor – 16, 17, 28, 34, 50, 53, 107, 108, 114, 126, 127
Project Gutenberg – 34, 36, 54, 59, 64
Public Domain – 32, 36, 37, 56, 59, 64, 80, 89, 100

Q

QuickTime – 15, 25, 79, 82, 83, 89, 90, 91–94, 119

R

RAM – 22, 28, 74, 127
Read Out Loud – 34, 67
ReaderWorks – 25, 51, 52, 120
Real Player – 28, 69, 77, 92
RealMedia – 93, 120
RealOne Player – 28, 69, 77, 92
Recorded Books – 59, 65
Resco Pocket Radio – 68
resolution – 15, 27, 29, 35, 89, 127, 1288
RIAA (Recording Industry Association of America) – 74
Riding the Bullet – 32
ripping – 65, 124
Rocketbook – 38
ROKR – 14, 15, 123, 127
Royalty Free – 73, 74
RSS feed – 46, 70, 74, 78, 80, 84, 85, 90, 94, 95, 127
RSS valuator – 78, 127

S

scientific probes – 15, 19, 20, 107
search engines – 40, 42, 80, 89
Secure Digital (SD) – 18, 19, 22, 56, 90, 109, 127
Simpson, Carol – 100
SingingFish – 89
Skype – 73, 127
smartphones – 14, 29, 59, 90, 92, 116, 121, 127
Soft Reset – 71
Software and Information Industry Association (SIIA) – 33
Sony Memory Stick – 18, 127
Sony PlayStation Portable – 15
   See also Sony PSP
Sony PSP – 93
SoundFlower – 79
Standards for the English Language Arts – 2, 124
Storyspace – 32
Symbian OS – 25, 127
Sync & Go – 46
synchronize – 8, 14, 18, 22–24, 27, 35, 43, 45–47, 50, 53, 56, 76, 78, 80, 81, 87, 91, 106, 113, 116, 125, 128

T

TEACH Act – 100, 122
TextAloud 67, 108
Text-to-Speech (TTS) – 25, 34, 66, 67, 78, 105, 106, 108, 128
TiVoToGo – 29, 90
Toshiba – 29
Transcriber – 16, 128

U

Unitedstreaming – 88, 109
U.S. Department of Education – 23, 40, 124

V

vidcast – 87, 128
VideoList – 113
Vincent, Tony – 71
Visual Artists Rights Act – 98
VITO SoundExplorer – 28, 67
VITO Voice2Go – 67
Vlog It – 95
Vodcast – xiii, 71, 87, 88, 94, 95, 98, 121, 124, 128
voice recorder – 17, 65, 66, 122

W

Wapedia – 43
Web clipping software – xii, 45, 123
WebCopier – 46
WiFi – 16, 19, 22, 67, 68, 73, 128
Wikibooks – 37
WinAmp – 59, 79, 81, 82
Windows CE – 14, 25, 26, 29
Windows Media Player – 14, 26, 27, 29, 59, 65, 68, 69, 76, 77, 81, 91, 92, 124
Windows Mobile – 14–16, 25, 27, 29, 34, 35, 41, 45, 46, 56, 78, 90–93, 112, 113, 123–128
Winer, Dave – 70
wireless LAN – 19, 20, 1288
Word – 19, 22, 24, 25, 34, 35, 36, 50–54, 62, 106, 116
WordSmith – 53

Y

Yahoo – 42, 72, 80, 89, 113

Index 133

www.ingramcontent.com/pod-product-compliance
Lightning Source LLC
Chambersburg PA
CBHW080412300426
44113CB00015B/2492